BUILDING YOUR LIFE SCIENCE VOCABULARY

BRYAN T. WHITWORTH

Respectfully dedicated to
Professor Wilbert J. Robinson
mentor and friend,
who taught me to know and love science

Wadsworth Publishing Company
I(T)P™ An International Thomson Publishing Company

Belmont • Albany • Bonn • Boston • Cincinnati • Detroit • London • Madrid • Melbourne
Mexico City • New York • Paris • San Francisco • Singapore • Tokyo • Toronto • Washington

Printed in the United States of America
1 2 3 4 5 6 7 8 9 10—01 00 99 98 97 96 95

For more information, contact Wadsworth Publishing Company.

Wadsworth Publishing Company
10 Davis Drive
Belmont, California 94002, USA

International Thomson Publishing Europe
Berkshire House 168-173
High Holborn
London, WC1V 7AA, England

Thomas Nelson Australia
102 Dodds Street
South Melbourne 3205
Victoria, Australia

Nelson Canada
1120 Birchmount Road
Scarborough, Ontario
Canada M1K 5G4

International Thomson Editores
Campos Eliseos 385, Piso 7
Col. Polanco
11560 México D.F. México

International Thomson Publishing GmbH
Königswinterer Strasse 418
53227 Bonn, Germany

International Thomson Publishing Asia
221 Henderson Road
#05-10 Henderson Building
Singapore 0315

International Thomson Publishing Japan
Hirakawacho Kyowa Building, 3F
2-2-1 Hirakawacho
Chiyoda-ku, Tokyo 102, Japan

Printer: Malloy Lithographing, Inc.

ISBN 0-534-26214-7

An Introduction to the Words Used in Science

If you have ever been confused or intimidated by the words used in science, this book is meant for you. Like most people, you have probably wished that scientists, science books, and science teachers wouldn't use such complicated words but would use common, everyday words instead. Here are some common words we use almost every day: automobile, photograph, helicopter, and television. These words not only sound right, they make sense.

One should realize, however, that these four words would surely have mystified the Americans of George Washington's day because all four of them are of Greco-Latin extraction, and at one time they were hard to understand. (If you would like to change them into English, the translations would be: automobile = *self-mover* ; photograph = *light-write*; helicopter = *screw-wing*; television = *distance-see*.) These words have become easy for us to understand only because we are familiar with them.

Long ago, scientists realized that they didn't have time to learn all the languages in the world; in order to communicate with one another, they decided to use Greek and Latin as common languages because at that time most educated people could read both of those classical languages.

Because of that decision, almost everything scientific employs Greek or Latin terms. Therefore, if one knows a few of the more common word elements in these two languages, the study of science becomes much easier.

To use this book to help you understand the real meaning of a science word, try to break the word down into elements or pieces. Then find the meaning of each element by looking it up in the pages that follow this introduction. You will find the elements (root, prefix, or suffix) listed at the left, the pronunciation next, the meanings near the center of the page, and some examples of usage of the element at the right side of the page.

Suppose that while reading a book on medical techniques you come upon the word aseptic. It begins with an a (an element that means *no*, *not*, or *without*) so you know the word means *without* the next part, septic. When you look up the meaning of septic, you find that it means *rotted.* Therefore, aseptic means *without rot.* A septic tank smells rotten, but something that is aseptic is clean (because aseptic things have no bacteria to cause rot). It takes little imagination to see that something that is antiseptic, would be a thing that works *against* (or kills) bacteria.

Let's take our word root A (meaning *not*) and combine it with the element theo (meaning *God*) to get the meaning *No God*. One who doesn't believe in God is an atheist. However, one who has no belief either way is an agnostic, because gnosis means *belief* (as in prognosis, which is a physician's belief about the outcome of a case after a diagnosis has been made).

Now let's take the root arch (meaning *first*, or *chief*) and combine it with mono, meaning *one* or *alone*: we get the word for a king—monarch. Arch also appears in anarchy, where it means *nobody first*: No person is in sole command. That might sound a little like democracy; but demos means *people*, and crat means *rule*—so democracy means *rule by the people*. Since demos means *people*, what does an epidemic mean? The alphabetized listings in this book show that epi means *on* or *upon*, which means that an epidemic is a thing (such as a disease or a fad) that is *on the people*.

By now you should be able to see that almost any element can be combined with another element to get a different meaning. For examples, auto means *self*. Combine it with mobile and you have the word automobile, a *self-mover*. Combine auto with eros (*love*) and you have auto-eroticism—one who sexually excites himself or herself. Combine auto with bio (meaning *life*) and then attach graphy (which means *description*) and you will see that autobiography means *self-life-description*. If you had left out the root bio, you would have an autograph, or *self-write*. On the other hand, if you combine gram (*written*) with the element mono (*alone*) you have a monogram—an ornate initial decorating a shirt, purse, handkerchief, or ring.

Graph, combined with tele (*long*, or *distance*) really means *long-distance-write*—but it's much easier to say telegraph! The element graph is also used with the element topo (*surface*) to get topography, which tells us what an area's surface is like; a topographical map shows the surface of the earth's mountains and valleys. Put the element geo, with graphy and you have geography, a writing about the earth.

The root homo means *same*. Combine this with geny (to *make*) and you get homogenized—wherein each drop of milk is the same as any other drop (as far as cream content is concerned). A homosexual is a person who prefers to have sexual relations with the *same sex*; a heterosexual prefers a *different-sex* partner. If we combine homo with zygo (meaning *yoke*) we have the word homozygous—meaning that an animal or plant had parents that were the same as far as any genetic trait is concerned. One can now easily see that a heterozygous creature had parents who were *different* in genetic makeup.

In the last paragraph, we saw that gen means *to make*. You already knew that a generator makes electricity, that a generation of people is a whole new population, and that genes are the stuff of life. Now you also know why the first part of the Bible is called Genesis.

Let us turn to the commonly used root logo (which means *speech*). If logo is combined with dia (meaning *across*) we get dialogue, which (in English) means *talk between* or *talk across*. When logo is used with the prefix pro (meaning *before*) we have the word prologue—for an introductory talk. But when logo is combined with the element eu (for *good*) we have eulogy—speech that is given at funerals as well as on happier occasions.

Logo, combined with mono, gives us monologue (where one person does all the talking). And mono, combined with graph, makes the word monograph—for something written about just one thing...such as science words.

Logy (usually appearing as ology) means *science* or *the study of*. Ology can be combined with many other elements to form a surprising number of words. Put it with ornithos (*bird*) and you have ornithologist—a scientist who studies birds. Ology also fits well with icthyo (*fish*) to denote an ichthyologist—one who studies fish. A geologist studies the earth, entomologists study insects, and ecologists study the interrelationships of all things on our home, the earth...because ecos means *home*.

Science words have real meanings. If you break down the word for a "bug doctor" (entomologist) into its elements, you have en = *inside*; tomo = *cut into*; and ology = *study*. An entomologist is a scientist who studies creatures that *cut into things*.

The element ology used with astr (*star*) once described those who studied the stars—astrologists. But modern scientists who study the heavens have cast the word aside because of the unfounded ancient belief that stars influence human endeavors. (Have you ever heard someone say "You can thank your lucky stars"?) Nowadays, those who study the stars prefer to be called astronomers, which means *starnamers* or *stargovernors*.

Consider the element astr, which combined with nautic (or *sailor*), means *one who sails to the stars*—an astronaut. Does that mean that Americans sail only to the stars while the Russian cosmonauts sailed to the whole *universe*? (Cosm means *universe*.)

By now, it should be obvious that the same element may be used in many different words that carry many different meanings. And it should also be obvious that you have been speaking (and reading) Greek and Latin words most of your life...perhaps without realizing it.

This book is an outgrowth of my years as a student, an entomologist, and a science teacher. Certain parts are excerpts from my radio program, *Let's Talk About Words*, but most of the work grew from the notes I developed to help my own students better understand some of the difficult words used in the biological sciences. I believe you will be pleased to see how many "hard" words will turn out to be easy when you use this book to tie together what you already know.

When using this book to understand words that seem difficult, first look at the word's root(s). Then look through the list of alphabetized roots. The meaning of the element will be near the page's center, while some other words that also use that root will be shown as examples of usage.

If the word-root you seek is not listed, it may be from a word that is neither Greek *nor* Latin. Perfectly acceptable science words (of German, French, or Arabic origin) are not included here because this book contains only words that are of Greek or Latin extraction.

Many of the elements given here might seem to have no immediate value. Some you may never use, but the chances are good that as a student of science you will come across them from time to time. Although there are many hundreds of word elements in the pages that follow, no attempt has been made to include all the possible elements. Some elements are so rare that only a specialist in certain branches of science might ever need them; others may apply to only one or two science words. Still others are so common in English that there is no need to include them here—their meaning is evident to anyone who speaks our language.

You may notice that there are no science words which begin with W or Y. The Greeks and Romans did not use these two letters; that is why no elements begin with a W or a Y in this book.

Etymologists and students of the classical languages might wish to argue my usage (and/or spelling) of certain elements. Let them! This book was written for people who are studying science. Rest assured that all the elements' spellings (as well as their meanings) agree with present American scientific usage.

You will notice that dashes appear next to most of the roots. Those dashes indicate the element's position in a compound word. Two dashes before an element mean that the element is normally a suffix that will appear at the end of a compound word. Dashes at the element's right mean that the element is normally a prefix in a compound word. Dashes on *both* sides of an element mean that it may be used as a prefix *or* as a suffix. No dashes at all indicate that the element is the main part of a scientific word and can usually stand alone. (The absence of dashes can also mean the element is normally used as a combining word—in the middle of a compound word.)

For every element in the lists that follow, there are at least three examples of words in which that element is used in some way and (because one of the purposes of this book is to show the students that they already know many scientific word-parts) a commonly known word is included in those examples whenever possible. You may notice that the examples are not in alphabetical order; the order in which the examples appear has been selected to reassure readers of their own knowledge. For example, sidus is Latin for *star.* Many people know that sidereal time is a method of setting earthly clocks by noting the position of the stars. People also know that chile con carne means chili *with* meat; but not many people have reasoned that our common English word consider means "*with stars.*" (The ancients wanted to act when the stars were with them, not against them.)

The rather wide spacing in the word lists has been used to allow the reader to insert any additional word elements where they are needed. The wide spacing also makes it easier to follow the pronunciation guide and examples of usage.

Because some instructors may disagree about the way to pronounce a few scientific terms, *there is a long and honorable tradition that all students should pronounce those words in the way their current teacher does.*

Accents vary by region. There was no agreement on what the "proper" accent should be until about 1939, when it was agreed that the pronunciation of words in the United States should approximate that of the majority of radio announcers *of that era*. That is the pronunciation given here but, because the pronunciation guides in some dictionaries are so vague (and the guides in others are so complex that only a few lexicographers can interpret them), each of the elements and their examples are spelled phonetically. The CAPITalized syllables receive vocal stress. The Short "i" (as in the word stiff) is written "ih" and is pronounced with a slight breathiness as in unicorn (which is phonetically written as YOUnihcorn). The open short vowel “e” (as in endemic and elude) is usually written as “eh”—which means those two words are phonetically written, in this book, as ehnDEMik and ehLOOD. The long “o” (as in amino) is written as “oh”; it may also appear as “-o-” when used as a combining form but, if the meaning might not be clear, the long “o” may have a line atop it as in myKŌsis. The short “o” will be written as “ah” to phonetically spell dichotomy as dyeKAHTohmee. Whenever possible a well-known word (or English name) will be used, as in FLEEbo for phlebo, NEEDoh for cnido, or dye-ALICE'S for dialysis.

ROOT/PREFIX/SUFFIX	PRONUNCIATION	MEANING	EXAMPLES OF USAGE
A--	A, or Ā (as in "hay")	No, Not	Agnostic, Atheist, Asexual, Anarchy
AB--	AB (as in "about")	From, Away	Abjure, Absorb, Abscond, Abstract
ABAT--	ahBAIT	to Beat Down	Abate, Abatable, Abatement
ABIO--	Ā BUYoh	Lifeless	Abiochemistry, Abiogenesis, Abiosis
ABYSS--	Abbess	Deep Bottom	Abyssal, Abysmal, Abyssic
ACA--	AHkuh	Nettle, Point	Acaleph, Acanaceous, Acanthus, Acanthoid
ACET--	uhSET, ahSEET	Sour, Vinegar	Acerbic, Acetabulum, Acetic acid
ACIN--	ASSin	Sword, Scimitar	Aciform, Acinaceous, Acinarius
ACINO--	AHseeno	Full, Grapes	Acinous, Aciniform, Acinus
ACOU--	ahCOO, AKyou	Hear	Acoumeter, Acoustics, Acoustician
ACRI--	ACKree	Sharp, Sour	Acrid, Acrimonious, Acrimony
ACRO--	ACKrow	Height, Extremity	Acrobat, Acrodont, Acromegaly
ACTIN--	ACKtin	Light Ray, Beam	Actiniform, Actinic, Actinozonal, Actinolite
ACU--	ACKyou	Needle, Sharp	Acupuncture, Acute, Acuity, Acumen
--AD--	Add	On, To(ward)	Addict, Adsorb, Adherent
ADENO--	ADDin-oh, ahDEENoh	Gland	Adenoid, Adenoma, Adenovirus
ADIPO--	ADDYpo	Fat	Adipoma, Adipose tissue, Adipocere
ADNAT, ADNAC--	ADnat, ADnas	Growing Together, Attached to	Adnascent, Adnate, Adnation

ROOT/PREFIX/SUFFIX	PRONUNCIATION	MEANING	EXAMPLES OF USAGE
AERI, AERO--	AIRy, Ā-EAR-oh	Air, Gas	Aerenchyma, Aeriferous, Aerial, Aerobic
ALA--	AHla	Wing	Aileron, Ala, Alate, Alar
ALBU--	ALboo	White	Albumin, Albuminous, Albuminuria
--ALGIA	ALjee-ah	Pain	Neuralgia, Cephalgia, Oralgia
ALIM--	ALim	to Nourish	Aliment, Alimentary canal, Alimony
ALLO--	ALoh	Different, Other	Allogamy, Allomorph, Allocate
ALLUV	ahLOOV	to Wash away	Alluvial, Alluvion, Alluvium
ALT--	ALt	High	Altitude, Altisonant, Altimeter, Altocumulus
ALVEO	ALVEEoh	Cavity	Alveary, Alveolar, Alveolus, Alveoli
AMALGA	ahMALga	Soft Material	Amalgam, Amalgamate, Amalgamative
AMAT--	AMmut	Sexual Love	Amative, Amatory, Amativeness, Amateur
AMBI--	AMbee	Both	Ambidextrous, Ambivalence, Ambiguous
AMBLY--	AMBlee	Obtuse, Blunt	Amblyopia, Amblygonite, Amblyoform
AMBU--	AMByoo	Walk about	Ambulacrum, Ambulance, Ambulatory
AMIN--	AHMeen	from Ammonia	Amine, Amino acid, Ammoniac
AMNIO--	AMnee-oh	Lamb	Amnion, Amniotic, Amniota
AMPHI--	AMfee	Both, Two Kinds	Amphibious, Amphineustic, Amphipod
AMPLI--	AMPlih	to Increase	Amplicative, Amplify, Amplitude, Ample

ROOT/PREFIX/SUFFIX	PRONUNCIATION	MEANING	EXAMPLES OF USAGE
AMYGDA--	ahMIGda	Almond	Amygdala, Amygdaloid, Amygdalin
AMYL--	AMil	Starch	Amyl alcohol, Amylopsin, Amylase
AN--	ANN	No, Not	Anarchy, Ananthous, Aneroid
ANA--	ANNA	Up, Again	Anaphase, Analogue, Analogy
ANAE, ANE--	ANNIE, ANNeh	without Feeling	Anesthetic, Anesthesia, Anesthesiology
ANEMO--	ahNEEmo	Wind	Anemone, Anemometer, Anemophilous
ANCO--	ank-o-	Bend	Ancon, Anconal, Ancoral, Anchor
--ANDRO--	ANNdro	Man, Male	Gynandromorph, Androclinium, Android, Androspore
--ANE	ĀNE	Attribute	Mundane, Arcane, Germane, Humane, Urbane
--ANE	ĀNE	Position (in organic chemistry)	Butane, Ethane, Methane, Octane, Propane
ANGINA--	anJINAH	Pain, Torture	Angina Pectoris, Anginoid, Anginal
ANGIO--	ANJEE-oh	Cased, Closed	Angiosperm, Angioblast, Angiocarp
ANGU--	ANN-goo	Having narrow Angles	Angulate, Angularly, Angustate
ANHYDR--	ANN-hider	free of Water	Anhydrate, Anhydride, Anhydrous ammonia
ANIM--	AHneem	Soul, Breath, Spirit	Animate, Animal, Animaculum
ANK--	ANNk	a Hook	Anker, Ankylose, Ankylosis, Anchor

ROOT/PREFIX/SUFFIX	PRONUNCIATION	MEANING	EXAMPLES OF USAGE
ANNEL--	ANNuhl	Year, Ring	Annual, Annelid, Anniversary
ANO--	ANNoh	Up, Absent	Anonymous, Anorexia, Anosmia, Anode
ANORTH--	ahNORTH	Up Straight	Anorthite, Anorthoscope, Anorthosite
ANSA	ANSuh	a Handle	Ansated, Inansate, Ansate cross
ANTE--	ANtay or ANNteh	Before	Antechamber, Antemortem, Ante meridiem
ANTH--	ANth	Flower	Ananthous, Antheridium, Antherozoid
ANTHRA--	ANTHrah	Coal	Anthrax, Anthracite, Anthracosis
--ANTHROP--	ANTHrup	Human, Mankind	Philanthropy, Anthropology, Anthropoid
ANTI--	ANtih, ANTeye	Against	Antidote, Antiseptic, Anticlimax
ANTIQ--	ANTik, annTEEK	Ancient	Antiquary, Antique, Antiquarian
APER--	Appear	Open	Aperient, Apert, Aperture, Aperitif
APHA, APHE--	AFfa, AFFeh	Dark, Obscure	Aphanite, Aphelion, Apheliotrophic
API--	Apih, APee	Top, Apex	Apical, Apex, Apiculate
APO--	APo, AFo	Off, Away from	Apotropaic, Apology, Apospory, Apogee
APOPH--	aPOFF	to Flee from	Apophasis, Apophthegm, Apophyge
APPENDI--	aPENDY	Addition, Hang on	Appendix, Appendicle, Appendage
APPET--	APPet	to Seek	Appetency, Appetite, Appetitive
AQUA--	AHKwa	Water	Aquanaut, Aquatic, Aqueduct

ROOT/PREFIX/SUFFIX	PRONUNCIATION	MEANING	EXAMPLES OF USAGE
ARACHN--	ahRAKIN	Spider	Arachnid, Arachnoid, Arachnology
ARBOR--	ARBer	Tree, Latticework	Arboreal, Arboretum, Arborescent
--ARCH--	ARK	First, Chief	Monarch, Archaic, Patriarch
ARCHI--	ARKih, ARTshe	Ancestral	Archiblast, Archive, Archenteron
ARGENT--	ARJent	Silver	Argent, Argentiferous, Argentine
ARTHRO--	ARTH-row	Jointed	Arthritis, Arthropod, Arthralgia
ASCO--	ASKo	Sack, Bladder	Ascocarp, Ascogonium, Ascomycete
--ASTER--	ASTr	Star	Astronomy, Asteroid, Astronaut, Disaster
ATMOS--	ATmos, ATmus	Vapor	Atmosphere, Atmolysis, Atmometer
AUDI--	AWEd	Hear	Audible, Audio, Auditorium
AURI--	OReee	Ear	Auricle, Auricular, Aurilave
AURU--	ORRoo	Gold	Aurora borealis, Aurum, Aureate
AUTO--	AWE-toe	Self	Automobile, Autonomy, Autobiography
AUX--	AWKS	Grow	Auxin, Auxiliary, Auxospore
AVIS--	AVE-iss	Bird	Aviator, Aviary, Aviation
AXI--	AKSih	Axis, Turn	Axilla, Axillary, Axil, Axle
AZA--	AZah	Nitrogen	Azine, Azole, Azonium
BAC--	BAK, BAS	Rod, Stick	Bacillus, Bacteria, Bacteriocide
BARI--	BARE-ih	Heavy	Barometer, Barium, Baritone

ROOT/PREFIX/SUFFIX	PRONUNCIATION	MEANING	EXAMPLES OF USAGE
--BASIS	BASEiss	Rhythmical movement	Abasis, Abasic, Abasia
--BASSUS--	BASE-us	Reduce	Abase, Debase, Bass, Base
BATHOS--	BATHoze	Deep	Bathysphere, Batholith, Bathymeter
BATUO--	BATwō	Beat down	Abate, Abatement, Abatable
BENTHO---	BENTHoh	Depths (of the sea)	Benthic, Benthal, Benthos
BI--	BUY	Two	Binary, Binomial, Bicycle
BIB--	BIB, BIBE	Drink	Imbiber, Bibulous, Bibacty
BIBL--	BIBl, BUYbl	Book	Bible, Bibliophile, Bibliography
BIO--	BUY-oh	Life	Biology, Biography, Biochemistry
--BLASTO--	BLAStoe	Embryo, Germ	Blastoderm, Blastocoel, Blastocyst, Ameloblast
--BOLI--	BOWLee	Ball, Throw	Bolide, Catabolism, Metabolism
BOS, BOV--	Boss, BOVE	Ox-like	Bovidae, Bovinae, Bovoid
--BRACHI--	BRAKEee	Arm	Brachial, Embrace, Brachiopod, Brachium
BRACT--	BRAKT	Thin Metal Plate, Modified Leaf	Braconnier, Bracteate, Bractiform
--BRANCHI--	BRANkee	Gill	Branchiomorph, Branchiopod, Elasmobranch
BREVI--	BREVih	Short	Brief, Brevilineal, *Yucca brevifolia*
BRONCHI--	BRAWNkee	Windpipe	Bronchitis, Bronchoscope, Bronchial
BRONT--	BRAHNT, BRŌNT	Thunder	Brontograph, Brontosaurus, Isobront

ROOT/PREFIX/SUFFIX	PRONUNCIATION	MEANING	EXAMPLES OF USAGE
BRYO--	BRĪ-oh	Moss, to Swell	Bryophyte, Bryology, Bryozoa
BU, BO--	BOO, BO	Ox	Buffalo, Boophilus, Bovine, Bucephalus
BUBO--	BOOBoh	Groin	Bubo, Bubonic, Bubonocele
BUCCA--	BOOkah	Cheek, Mouth	Buccal, Buccolingual, Buccinator
BULLA--	BOOLah	of Lead	Bullet, Bulletin, Papal bull
BUNIO--	BOONyo	Hill, Rounded	Bunion, Bunodont, Buninoid
--BURSA--	BURSah	Purse, Pouch	Disburse, Bursitis, Bursiform, Reimburse
BUTY--	Beauty	Butter	Butyl, Butyric, Butyrin
BYSS--	BISS	Flax, Fiber	Byssaceous, Byssoid, Byssus
CADUC--	kaDOOS	Dropping off	Caduceus, Caducity, Caducigen
CALC--	KALSS	Lime	Calciphilous, Calcite, Calcium
CALCI--	KALSEE	Chalk, Lime	Calcification, Calcify, Calcimine
CALCU--	KALkew, KALcoo	to Reckon, Pebble	Calculus, Calculate, Calculator
CALE--	KALeh	Warm	Califacient, Calefactory, Calify, Calorie
CALIC--	KALik	Cup	Calix, Calicle, Caliculus
CALLI--	KALee	Beauty	Calliphoridae, Callitroga, Callimania
CALO--	KALLO	Heat	Calorie, Caloreceptor, Calorescence
CALYP--	KAYlip	Husk, Cover	Calyx, Calypteron, Calyptrogen
CAND--	Canned	White, Bright	Candle, Candescent, Candent

ROOT/PREFIX/SUFFIX	PRONUNCIATION	MEANING	EXAMPLES OF USAGE
CANDI--	Candy	White, Shine	Candid, Candor, Candidate
CANIS	KANE-iss	Dog	Canine, Canicular, Kennel
CANNA--	KANna	Reed	Canister, Cane, Cannula
CAPI--	KAPih	Hair	Capillarity, Capillitium, Capillary
CAPRIC--	KAPrik	Goat	Capriolate, Caper, Capriform
CARBO--	KARbo	Coal	Carbohydrate, Carbon, Carbuncle
CARCIN--	KARsin	Cancer	Carcinoma, Carcinogen, Carcinoid
--CARDI--	KARDih	Heart	Mesocardium, Cardiac, Cardiogram
CARNA, CARNI--	KARNa, KARNih	Flesh	Carnivorous, Carnal, Carnival
CARO--	KARoh	Flesh, Carrion	Carbuncle, Carunculous, Carophagous
--CARPO--	KARpo	Wrist	Metacarpal, Carpus, Carpal
CARPOS--	KARPos	Fruit	Carpology, Carpogonium, Carpophore
CARYO--	KARRY-oh	Nut	Caryocar, Caryopsis, Caryota
CASE--	KASE	Cheese	Caseate, Casein, Caseinogen, Caseation
CAST--	KAST	to Make Pure	Castigate, Caste, Cast
CASTOR	KAStor	Beaver	Castoreum, Castor oil, Castor bean
CATA--	KATah	Down, Against, or Entirely	Catalogue, Cataclysm, Catastrophe
CATEN--	kahTEEN	a Flexible Chain	Catena, Catenate, Catenulate
CATHAR--	KATHer	Pure, Purge	Catharsis, Cathartic, Catheter, Catharist

ROOT/PREFIX/SUFFIX	PRONUNCIATION	MEANING	EXAMPLES OF USAGE
CAUDA	KAWdah	Tail	Caudal fin, Acaudal, Caudate
CAULI--	KAWli	Stem	Caulescent, Caulicle, Cauliflower
--CAUST--	KAWST	Burnt	Caustic, Holocaust, Cauterize
CAV--	Kāve	Hollow	Cave, Concave, Caval, Cavern
CAVAL--	KAVul	Horse	Cavalier, Cavalcade, Cavalry
--CEDO	SEEDoh	Go	Proceed, Precede, Exceed, Cedula
CEIL--	SEAL	Heaven	Ceiling, Celestial, Celestite
CELL	SELL	Close Room	Cell, Cellar, Cellarette, Cellular
CEMENT	siMENT	(of things finely) Cut	Cementer, Cementation, Cementite
CENO--	SEEN-oh	Empty	Cenotaph, Cenosis, Cenotron
CENO--	SEEN-oh	New, Common	Cenozoic, Cenobite, Cenosere
CENS--	SENz	to Judge	Censorial, Censor, Censure
CENT--	SENT	Hundred	Centigrade, Century, Centimeter, Percent
--CENTR--	SENTer	Round, Middle	Eccentric, Ethnocentric, Centromere
CENTRO--	SENTrō	Center	Centerpiece, Central, Centering
CEPHAL--	SEFFul	Head	Cephalic, Encephalitis, Cephalothorax, Cephalopod
--CEPS	SEPS	Head	Biceps, Triceps, Concept
--CEPT	SEPT	to Take	Reception, Intercept, Percept

ROOT/PREFIX/SUFFIX	PRONUNCIATION	MEANING	EXAMPLES OF USAGE
--CERAT--	SERHut	Horn	Rhinoceros, Cerargyrite, Keratin
--CERC--	SURK or SURSS	Tail	Cerci, Cercopidae, Heterocercal
CEREB--	sirREEB	Brain	Cerebrum, Cerebral, Cerebellum
CERES	SĒAReez	Corn, Grain	Cereal, Cerealla, Cerealin, Cerealine
CERO--	SEHRoh	Wax	Cerolein, Cerotic, Cerotype
CERVI--	SERVih	Neck	Cervix uteri, Cervical, Cervix
--CHAETO--	KEY-toe	Hair, Bristle	Chaetognath, Chaetopod, Chaetophorus, Spirochete
CHELA--	KEY-lah	Claw	Chelate, Cheliferous, Cheliform
CHLOR--	KLŌRE	Green	Chlorophyll, Chlorine, Chloroplast
CHOL--	KŌLE	Bile, Gall	Cholesterol, Cholera, Cholinesterase
--CHONDR--	KAWNder	Cartilage	Hypochondria, Mitochondria, Chondriosome
--CHORD--	KORD	Cord, String	Chordata, Notochord, Chordate
--CHRIN	KRIN (like "grin")	Separate	Endocrine, Endocrinology, Exocrine
CHROM--	KRŌME	Color	Chromosome, Chromophore, Chromium
CHRONO--	KRŌ-no	Time	Chronological, Chronometer, Anachronism
CHRYS--	KRISS	Gold	Chrysalis, Chrysaniline, Chrysochlor

ROOT/PREFIX/SUFFIX	PRONUNCIATION	MEANING	EXAMPLES OF USAGE
--CHYM--	KĪME	Juice, Infusion	Parenchyma, Sclerenchyma, Chyme, Chymosin
--CIDE	SIDE	to Kill	Insecticide, Suicide, Homicide
CILIA--	SILLYah	Little Hairs	Ciliata, Cilium, Ciliograde
CINE--	SEENeh or SINeh	Moving	Cinema, Cinematic, Cinemascope
CIR--	SEER	Curl, Tendril	Cirrus, Cirrocumulus, Cirriped
CIRCUM--	SURKum	Around	Circumflex, Circumference, Circus
--CISE--	SĪSE or SIS	to Cut	Incise, Incision, Scissors, Circumcision
CLASS--	KLASS	Clan, in Common	Classify, Classification, Classmate
--CLAST--	KLAST	to Break	Iconoclast, Orthoclase, Plagioclastic
CLAUST--	KLAWST or KLAH-ōst	to Close, Shut	Claustrophobia, Closet, Clausura
CLAV--	KLAV (rhymes with "have")	Key, Club	Clavicle, Clavichord, Clavate
CLEISTO--	KLĪSToh	Closed	Cleistocarp, Cleistogamous, Cleistomorph
CLIMA--	KLEEMah or CLĪMah	Region, Slope	Climate, Climax, Climacteric
--CLIN--	KLĪNE	to Lean	Incline, Clinograph, Recline
CLON--	KLAHN, KLŌNE	Twig, Motion	Clone, Clonic, Clonal
--CLUD--	CLOOD	Closing	Conclude, Occlusion, Preclude

ROOT/PREFIX/SUFFIX	PRONUNCIATION	MEANING	EXAMPLES OF USAGE
CNIDO--	NEEDō	Nettle	Cnidoblast, Cnidophore, Cnidocil
CO--	KŌ	With	Coed, Cohere, Cohabitate, Cohesion
CO--	KŌ	Needed for Completion	Codeclination, Cosecant, Cosine
COAGU--	koAHGyoo	Curdle, Gel	Coagulate, Coagulin, Coagulum
COALESC--	koahLESS	Nourish Together	Coalition, Coalesce, Coalescence
COCC--	KAHKS, KŌKES	Berry	Coccidiosis, Coccidium, Streptococcus
--COCH--	KŌKE (sometimes COACH)	Snail	Cochlea, Cochleant, Cochleated
--COCT--	Cocked	Cooked, Bake	Coctil, Concoction, Coction
--COEL--	Seal	Cavity	Coelocanth, Coelenterata, Cecum, Hydrocele
COGNI--	KOGnee (sometimes KŌNyee)	to Know	Recognize, Cognizant, Cognoscenti
COLEO--	KOLEEoh	Sheath	Coleophyl, Coleoptera, Coleorhiza
COLLO--	KŌL-lō	Glue	Collodion, Colloid, Collodium
COM--	KUM, KAHM	Together	Complicate, Compose, Complex
CON--	KŌNE	With	Conclude, Conjugate, Connect
CONA--	KŌNah	Effort, Attempt	Conation, Conatus, Conative verb
CONIDI--	kōNEED-o	Dust	Conidium, Conidiophore, Conidial

ROOT/PREFIX/SUFFIX	PRONUNCIATION	MEANING	EXAMPLES OF USAGE
CONJUNCT--	kunJUNKT	Join Together	Conjunctiva, Conjunctivitis, Conjunction
CONO--	KŌ-NŌ	Cone	Conodont, Conoid, Conoscope
CONSIDER	kunSIDDer	with Stars	Considerable, Considerate, Consideration
CONTRA--	KŌNtra	Against	Contrary, Contradict, Contraception
--COPIA--	KŌPEEah	Abundance	Copious, Copiously, Cornucopia
COPRO--	KŌPE-rō	Dung	Coprolite, Coprolallia, Coprophagous
CORNU--	KORNoo	Horn	Cornucopia, Cornea, Cornu
CORONA--	kaRŌNA	a Ring	Corrola, Coronary, Coroniform, Coronium
CORPUS--	KŌRpus	Body	Corpus luteum, Corpse, Corporation
CORRO--	KOR-o or ka-RŌ	Together-gnaw	Corrode, Corrosive, Corrodent
CORTEX	KORtex	Tree Bark	Corticate, Cortical, Cortin, Corticose
COSM--	KAHZM	Universe	Cosmonaut, Cosmic, Cosmopolitan
--COST--	Coast, or KAHST	Ribs	Intercostal, Costalgia, Costo-abdominal
--COTYL--	COATihl	Cup	Cotyledon, Monocotyledon, Dicotyledon
CRASS	KRASS	Thick	Crassis, Crassamentum, Crassulaceous, Crass
--CRAT	KRAT (as in “that”)	Strength, Rule	Autocrat, Democrat, Bureaucrat

ROOT/PREFIX/SUFFIX	PRONUNCIATION	MEANING	EXAMPLES OF USAGE
CREDO--	KREED-o	Believable	Credit, Incredible, Credulous
CREO--	KREE-o	Flesh	Creodont, Creatoxin, Creatorrhea, Creosote
CREPUSC--	krayPUSK or kreePUSK	Twilight	Crepuscular, Crepuscle, Crepusculum
--CRESC--	KRESK or KRESS	to Grow	Accrue, Crescent, Accrete, Accretion
CRIN--	KRIN (as in "grin")	Lily	Crinite, Crinoid, Crinoldia, Crinum
CRUCI--	KREWsee	Cross	Cruciform, Cruciate, Excruciating
CRYO--	CRY-o	Cold	Cryogenics, Cryostat, Cryology
CRYPT--	KRIPT (as in "script")	Secret, Hidden	Cryptic, Crypt, Cryptogamous
CTENO--	TEENoh	Comblike	Ctenidium, Ctenodean, Ctenophore
--CUBO--	KYOOBoh	Lie in (or on)	Incubator, Incubus, Incubate
CULI, CULE--	kYOOLee	Gnat	Culicidae, Culicoidal, Culicifuge
CUMU--	KYOOM-yoo	Pile, Heap	Accumulate, Cumulus, Cumulative
CUNE--	KEWNih	Wedge, Triangle	Cuneal, Cuneiform, Cuneate, Cunnilingus
CUPR--	KOOPr	Copper	Cuprum, Cuprite, Cuprous, Cupric
CURS--	Curse	Run	Cursorial, Cursive, Cursory, Precursor
CUSP	KUSp	a Point	Cuspid, Cusped, Cuspidate, Bicuspid
CUTI--	Cutie	Skin	Cutin, Subcutaneous, Cuticle
CYANO--	sighANN-o	Dark Blue	Cyanic, Cyanide, Cyanosis

ROOT/PREFIX/SUFFIX	PRONUNCIATION	MEANING	EXAMPLES OF USAGE
--CYCLE--	SIGH-kul	Circle	Bicycle, Cyclone, Cyclic, Unicycle
CYME--	SIME (as in "time")	Wave, a Shoot	Cymene, Cymograph, Cymose
CYRT--	SURT (sometimes KURT)	Bent, Bulging	Cyrtidae, Cyrtoceras, Cyrtometer
--CYST--	SISt	Bladder	Cystogram, Cystology, Cystorrhea, Nematocyst
--CYTO--	SITE-o	Cell	Cytology, Lymphocyte, Cytoblast
DACRY--	DACKree	Tears	Dacryocyst, Dacryogenic, Dacryolith
--DACTYL--	DACKTul	Finger	Pterodactyl, Dactylogram, Pentadactyl
DEC--	DECK or DESS	Ten	Decade, Decimal, Decagon, December
DECI--	DESSee	Falling out	Decimate, Deciduous, Decidua
DELTA	DELtah	Change, Triangular	Deltoid, Deltic, Deltoidal
--DEM--	DEHM	People, Populace	Epidemic, Democrat, Endemic
DENDR--	DENdr	Tree	Dendriform, Dendrite, Dendrolith
DENT--	DEHNT	Tooth	Dentist, Denticulate, Dentition
--DERM--	DURM	Skin	Hypodermic, Dermatology, Ectoderm
DEUT--	DEWt	Second	Deuthyalosome, Deuteronomy, Deutoplasm
DEXT--	DECKST	Toward the Right	Dexter, Dexterity, Ambidextrous
DI--	DYE or DI (as in "dim")	Apart	Dioecious, Digress, Diverge, Diastyle
DI--	DYE	Two, Twice	Dichromate, Diazine, Diptera, Dichotomy

ROOT/PREFIX/SUFFIX	PRONUNCIATION	MEANING	EXAMPLES OF USAGE
DIA--	DYE-ah	Across, Through	Diagram, Dialysis, Diaphragm, Diarrhea
--DIAZO--	dyeAZ-o	Two Nitrogens	Diazotize, Diazonium, Diazohydride
DICT--	DIKT	to Say	Dictum, Dictation, Dictator, Contradict
DIDACT--	DYE-DAKT	to Teach	Didactic, Didacticism, Didact
DIFFER	DIFFur	Carry Apart	Different, Differentiate, Difference
DIGIT	DIHJit	Finger, Five	Digitate, Digitalis, Digitiform
DILU--	DILLyoo	Wash Apart	Dilute, Diluent, Dilution
DINO--	DYE-no	Terrible	Dinocerate, Dinosaur, Dinothere
DIPLO--	DIP-low	Double	Diploid, Diplodicus, Diplophyl, Diplopia
DIVI--	DIVih	Cut, Apart	Divident, Division, Dividant, Dividend
--DONT--	DAHNt	Tooth	Orthodontist, Lophodont, Coelodont
DORM--	DORm	Sleep	Dormancy, Dormitory, Dormant
DORS--	DORs (as in "horse")	the Back	Dorsal, Dorsolateral, Dorsal fin
--DROME--	DROHm (as in "home")	Course, Run	Aerodrome, Hippodrome, Dromedary
--DUCT--	Ducked	to Lead	Conduct, Aqueduct, Ductile, Induct
--DURA--	DOOrah	Hard	Durable, Dura mater, Durain, Endure
DYN--	Dine (as in "wine")	Powerful, Able	Dynamite, Dynamic, Dynamo, Dynagraph
DYS--	DISS	Ill, Bad	Dysentery, Dyspepsia, Dyslexia, Dystaxia

ROOT/PREFIX/SUFFIX	PRONUNCIATION	MEANING	EXAMPLES OF USAGE
E--	EH, or EE	Out, Exclude	Emit, Edict, Elude, Erupt, Ebullient
EC--	EK (as in "heck")	Out of	Eccentric, Ecstasy, Ectopic pregnancy
ECHINO--	eeKĪNE-o	Spiney	Echinoderm, Echinus, Echinate
ECOS--	EHKōse	Home	Ecology, Ecotone, Ecosystem
ECTO--	EKTō	Outside	Ectoderm, Ectoplasm, Ectomorph
EFFERVESC--	EFFERvess	Out-boil	Effervescence, Effervescent, Effervescible
EFFU--	EHFFuh	Pour Forth	Effulgence, Effuse, Effusion, Effusive
EGO--	EEgo	Self	Egoism, Egoistic, Egotist, Egotistical
EJECT--	eeJEKT	Throw out	Ejection, Ejective, Ejecta, Ejectamenta
ELASMO--	eeLASSmo	Metal Plate, Gills	Elasmobranch, Elasmosaur, Elasmothere
ELAST--	eeLAST	Return to Shape	Elastance, Elastic, Elasticity, Elastically
ELAT--	ehLATE	Out-borne	Elated, Elater, Elaterin, Elaterium
ELECTR--	ehLEKTr	Amber	Electron, Electricity, Electrum, Electrify
ELUTR--	elYOOtr	to Wash Off	Elutriate, Elutriation, Elutrite
ELYTR--	ehLEETur	Sheath, Case	Elytron, Elytra, Elytroid, Elytrum
EMBOLI--	EHMbowl	(Throw a) Ball	Embolic, Embolism, Embolismic, Embolus
EMBRYO--	EMbree-oh	Swelling inside	Embryosis, Embryonic, Embryology

ROOT/PREFIX/SUFFIX	PRONUNCIATION	MEANING	EXAMPLES OF USAGE
EME--	eMEE	Vomit	Emesis, Emetic, Emetine, Antiemetic
EN--	EN (as in "hen")	In	Energy, Endemic, Encephalitis
ENANTIA--	eeNANshuh	Mirror-Image	Enantiomorph, Enantiomer, Enantiosome
ENCEPH--	ehnSEFF	in the Head	Encephalitis, Encephalitic, Encephalogram
ENDO--	END-o	Within	Endocrine, Endomorph, Endoskeleton
ENTERO--	ENTER-o	Intestine	Enterostomy, Enterobiasis, Enterobic
ENTOMO--	ENTOH-mo	Cut Into	Entomology, Entomophobia, Entomophilous
EO--	EEoh	Early	Eohippus, Eozoan, Eolithic, Eocene
EOS--	EEōss	Dawn, Rose Color	Eosere, Eosin, Eosinophil
EPHEME--	ehFEEM	Live Only a Day	Ephemeral, Ephemeridae, Ephemeron
EPI--	EPPee	Upon	Epidemic, Epitaph, Epithet, Epitome
EQUA--	EKwah	Equal	Equate, Equator, Equation, Equalization
ERG--	URG or EHRG	Work	Ergal, Ergonomics, Ergograph, Energy
EROS--	EHRōss	Love, Sexuality	Autoeroticism, Erotic, Erogenous
ERYTHRO--	ehREETHroe	Red	Erythrocyte, Erythromycin, Erythrite
ESO--	ESS-oh	Will Bear	Esophagus, Esostamenous, Esophoria
ESTIV--	EHstiv (as in "festive")	Summer	Estivate, Estival, Estivation

ROOT/PREFIX/SUFFIX	PRONUNCIATION	MEANING	EXAMPLES OF USAGE
ETHI--	ETHih	Character	Ethics, Ethically, Ethos, Ethicize
ETHNO--	ETHno	Race, Nation	Ethnocentric, Ethnology, Ethnic
ETYMO--	ETee-mō	the True (sense)	Etymology, Etymologist, Etymon
EU--	You	Good	Eugenics, Euphoria, Euthanasia
EUPHEM--	YOOfem	Good Report	Euphemism, Euphemistic, Euphemia
EURY--	YOORih	Wide	Aneurism, Eurypterid, Eurycephalic
EX--	EKS	Out of	Exodus, Exophthalmic, Exclude
EXO--	EKS-oh	Without	Exopathic, Exorhiza, Exopodite
FACI--	FASSih	Make (easy)	Facile, Facilitate, Facsimile
FACIES	FAYsees	Face	Facet, Facial, Faciend
FASCI--	FASSsih	Bundle of Rods	Fascia, Fascine, Fascist
FEBRI--	FEBreh	Fever	Febrifuge, Febricity, Febrile
FELI--	FELih	Happy	Felicitate, Felicity, Felicific
FELIS	FEELiss	Cat	Felidae, Feline, Felineness
FELS--	FELLss	Rock	Felsic, Feldspar, Felsite
FENESTR	fehNESTER	Window	Fenestela, Fenestrated, Fenestra
--FER--	Fehrr	Carry	Ferry, Conifer, Aquafer
FERA	FEHrah	Wild	Feral, Ferocious, Fierce
--FEROUS	FEHR-us	Bearing	Coniferous, Auriferous, Sporiferous
FERR--	Fehrr	Iron	Ferric, Ferromagnetic, Ferrous

ROOT/PREFIX/SUFFIX	PRONUNCIATION	MEANING	EXAMPLES OF USAGE
FIBRI--	FIBrih	Hairlike	Fiber, Fibrin, Fibroblast, Fibrinogen
FIDO	FEEDoh	Faithful, Trustful	Infidel, Fidelity, Fiduciary
FILO--	FILL-oh	Thread	Filopodium, Filament, Filarial
FISSI--	FISSee (as in "prissy")	Split, Cleft	Fissile, Fissure, Fissate
FLAGEL--	FLAJul (as in "agile")	Whip	Flagellum, Flagellate, Flagelliform
FLATU--	FLAToo	Blow, Break Wind	Flatulence, Deflate, Flatulate, Flatus
--FLECT	Flekt	Turn	Deflect, Reflect, Inflect
--FLEX--	Fleks	Bend	Flexor, Reflex, Flexure, Flexuose
FLOCC--	Flock	Lock of Wool	Floccillation, Floccose, Flocculent, Frock
FLOR--	Flōr	Bloom, Flower	Floret, Florescence, Florist
FLUOR--	Floor	Glow with Color	Fluoresce, Fluorescein, Fluorescence, Fluoroscope
--FLUVI--	FLOOvih	Flow, a River	Fluvial, Effluvium, Effluvial
--FLUX--	Flucks (sometimes Flukes)	Melt, Flow	Fluxion, Fluxional, Influx, Confluent
FOCA--	FŌK-ah	Focus, Point	Focal, Focometer, Focal infection
FOLI--	FOlee	Plant, Leaf	Folio, Foliage, Foliate
--FOLIO--	FOAL-yō	Leaves	Foliolate, Folium, Bifoliate, Trifoliate
FOLLI	Folly	a Sack, Bag	Follicle, Follicular, Folliculated
FOMENT	foMENT	Stir Up and Heat	Fomented, Fomentum, Fomentation, Fomenter

ROOT/PREFIX/SUFFIX	PRONUNCIATION	MEANING	EXAMPLES OF USAGE
FORAMI--	foRAMih	Opening, Hole	Foraminate, Foramen magnum, Foramen
--FORM--	Form	Shape	Fusiform, Vermiform, Formula, Pro forma
FOSSA--	FAWSah	Dig, Trench	Fossil, Fossorial, Fossilize
--FLICT--	Flicked	to Strike	Conflict, Inflict, Infliction, Inflictive
--FLU--	Flew	Channel, Flow	Influenza, Fluent, Confluence, Influence
--FRACT--	Frakt (as in "tracked")	Broken	Fraction, Fracture, Refraction, Infraction
FRAGI--	FRAJih	Delicate, Frail	Fragile, Fragment, Fragmental, Fragmentation
FRAT--	Frat (as in "rat")	Brother, Sibling	Fraternize, Fraternity, Fraternal twins
FRET, FRIC--	Frett, Frick	Rub, Carve	Frets, Fretful, Fricative, Friction
FRIGI--	FRIJih	Be Cold	Frigid, Refrigerate, Frigidity, Fritillary
FROND	Frahnd	to become Green	Frondiferous, Frondescent, Frondose, Fronds
FRONS	Frohnss or Frahnz (as in "lawns")	Forehead, Brow	Front, Frontogenesis, Frontolysis, Frontier
FRUC, FRUG--	Frook, Froog	Fruit	Fructose, Fructuous, Fructiform, Frugal
FRUST--	Frust (as in "rust")	to Fail Attainment	Frustrate, Frustom, Frustule, Frustration
--FUGI--	FEWjee	to Flee	Fugitive, Centrifugal, Fugacious
FULMIN--	FULLmen	Thunder, Brightly Gleam	Fulminant, Fulminating, Fulmineous, Fulminic
FUMA--	FOOMah	Smoke, Fumes	Fumarole, Fumigate, Fumaric acid
FUND--	FUNd	Bottom, Deep	Fundament, Founder, Fundus

ROOT/PREFIX/SUFFIX	PRONUNCIATION	MEANING	EXAMPLES OF USAGE
FUNG--	FUNje (more properly, FOONjee)	Mushroom, Toadstool	Fungus, Fungi, Fungicide, Fungal
FUNIC	FEWnik (more properly, FOOnick)	Small Rope	Funicular, Funiculate, Funiculus
FURCUL--	FURKyool, FURkool	Forked Prop	Fork, Bifurcate, Furcula, Furculum
FURI, FURO--	Fyoorih, FYOOR-oh	to Rave, Insane	Furor, Infuriate, Fury, Furious
FUSE	FYOOz	to Melt	Fusion, Fusible, Fuse, Infusible
FUSI--	FYOOsih	Spindle-shaped	Fusiform, Fusobacterium, Fuselage, Fusee
FUST	Fussed	a Club	Fustic, Fustigate, Fustate, Fusty
GALA--	GAYlah	Milk-white	Galaxy, Galactagog, Galactic
GALEA	GALeah, GHEELYah	Like a Helmet	Galeate, Galeated, Galeiform
GALLIN--	GALin'	Poultry, Cock	Gallinae, Gallinaceous, Gallinule, Gallinipper
GAMETO--	guMEEToh	Reproductive Germ Cell	Gamete, Gametocyte, Gametogenesis
GAMO--	GAMoh	United	Gamogenesis, Gamosepalous, Gamotropic
--GAMY	GYAMee, GAHMee	Marriage	Bigamy, Cryptogamy, Monogamy
GANGLI--	Gānglee	Tumor	Ganglia, Ganglion, Gangliated, Ganglionic
GASTR--	GAStur	Stomach	Gastroenteric, Gastrectomy, Gastritis
GEL	Jell	Freeze	Gelable, Gelatin, Reversible gel, Jelly
GEM--	Jem	a Bud	Gemma, Gemmoid, Gemmation, Gemmule

ROOT/PREFIX/SUFFIX	PRONUNCIATION	MEANING	EXAMPLES OF USAGE
GEMIN--	JEMmin	Twin	Gemel, Geminate, Gemination, Gimbal
GEN--	Jen	Birth, Make	Genetics, Genesis, Generator, Generation
GENIT--	JENit	Genitals, Beget	Genitals, Genitourinary, Genitalia, Genitive
GENO--	JEENoh, JENNoh	Race, Sex	Genotype, Genoblast, Homogenous, Genotypic
GENU--	JENyoo	Knee	Genuflect, Genuflection, Genuotomy
GENUS	JEEnus	to Become, Class	General, Generic, Genuine
--GENY	Jenny	Origin, Making	Nosogeny, Progeny, Ontogeny
GEO--	JEE-o	Earth	Geography, Geotropism, Geometry
GERI, GERO--	Jerry, JEH-rō	Old	Geriatrics, Gerontology, Gerontocracy
GERM	Jurm	Sprig, Seed	Germinate, Germicide, Germinal Disk
--GESTO--	JESToh	to Carry	Digestion, Gestation, Ingested
GIBBOS	GIBBōs	Hunched, Not Rounded	Gibbosity, Gibbous, Gibbousness
GIGA--	GIGah	Giant	Gigahertz, Gigameter, Giant
GLABR--	Glabbr	Without Hair	Glabrate, Glabrescent, Glabrirostral
GLACI--	GLAYsee	Ice	Glacial, Glaciate, Glacier, Glacis
GLADI--	GLADee	Sword	Gladiator, Gladiolus, Gladiolar
GLANS	Glanss	Acorn	Gland, Glandular, Glandule, glans penis

ROOT/PREFIX/SUFFIX	PRONUNCIATION	MEANING	EXAMPLES OF USAGE
GLAUCO--	glah-OO-kō	Blue-gray	Glaucoma, Glauconite, Glaucous
GLOB--	Glahb, Glōbe	Ball	Gamma globulin, Globerigina, Globin, Globe
GLOM--	GLAHM, GLŌME	Make a Ball	Glomerulus, Conglomerate, Glomerule
--GLOSSA--	GLOWsah, GLAHSSah	Tongue	Toxiglossa, Glossolalia, Glossary
--GLOTT	Glaht	Tongue	Polyglot, Glottal, Epiglottis
GLUCO--	GLUE-kō	Sweet	Glucose, Glucinum, Glucosamine
GLYCO--	GLAIkō	Sweet Oil	Glycerin, Hypoglycemia, Glycogen, Glycosuria
--GLYPH--	Gliff	Relief-carving	Petroglyph, Hieroglyphics, Glyptodont
--GNATH--	Nath (as in "lath")	Jaw	Prognathous, Gnathoplasty, Gnathopod
GNOMO--	Nō-mōs	Thought, Know	Gnomic, Gnomen, Gnomical projection
--GNOSIS--	NŌsis	Belief, to Know	Prognosis, Agnostic, Diagnosis, Gnostic
GON--	Gōn, Gahn	Sexual, Seed	Gonad, Gonorrhea, Gonophore
--GRAD--	Grade	Step	Centigrade, Graduate, Gradually, Gradus
--GRAM--	Gram (as in "ram")	Written	Telegram, Diagram, Grammar
GRAN--	Gran	cereal Grain	Granary, Granules, Graniferous, Grange
--GRAPH--	Graff	Write	Photograph, Topographical, Graphite

ROOT/PREFIX/SUFFIX	PRONUNCIATION	MEANING	EXAMPLES OF USAGE
--GRAPHY	GRAFee	Description	Geography, Enterography, Cardiography
GRAVID--	Gravid (as in "avid")	Heavy	Gravity, Gravid, Gravimetry, Grave
GULLA	GULLah	Throat, Red	Gules, Gullet, Gula, Gular
GUTTA--	GUTTah	Drops, Dripping	Guttate, Guttation, Gutter, Guttula
GYMNO--	JIM-no	Naked	Gymnasium, Gymnospore, Gymnosperm
--GYNE--	GUYneh	Woman	Gynecologist, Gynephobia, Misogynist
--GYRO--	JĪ-rō	to Whirl	Gyrate, Gyroidal, Gyroscope, Autogyro
HAEMO--	HEmō or HĀY-EEmō (HĀĒMO)	Blood	Hemostat, Hemoglobin, Hemolymph
HAL--	Hal or Hay	Salt	Halide, Halite, Halid, Halogen lights
HALLUC--	HalLOOS	Wander in Mind	Hallucinate, Hallucinogen, Hallucination
HAPL--	HAPel	Single, Simple	Haploid, Haplosis, Haplopia
HAPTO--	HAPtō	Fasten, Attach	Haptere, Haptophore, Haptophile
HEBDO	HEBdōugh	Week, Seven	Hebdomad, Hebdomadal, Hebdomadary, Hebdomary
HEBET--	HEBet	Dull, Stupid	Hebetate, Hebetation, Hebetude
HECTO--	Hektō	Hundred	Hectoliter, Hectare, Hectogram
HEGE--	Hej	to Lead	Hegemony, Hegemonic, Hegumenos

ROOT/PREFIX/SUFFIX	PRONUNCIATION	MEANING	EXAMPLES OF USAGE
HEIR, HERED--	Air, HEHRed	to Inherit	Heirloom, Heirless, Heir apparent, Heredity
HELI--	HEELih, HELLih	Spiral, Screw	Helix, Helicopter, Helicoid
HELIO--	HEELee-oh	Sun	Heliograph, Helium, Heliotrope
--HELMINTH--	HELLminth	a type of Worm	Helminthic, Helminthiasis, Platyhelminthes
HEMI--	HEMee	Half	Hemisphere, Hemialgia, Hemiptera
HEMO--	HEEmo	Blood	Hematite, Hemolymph, Hemolysis, Hematocele
HEPATO--	hehPAHToh	Liver	Hepatitis, Heparin, Hepatectomy
HEPTA--	HEPTah	Seven	Heptagram, Heptamerous, Heptahedral
HERB--	HURB (sometimes URB)	Grass, Herbage	Herbaceous, Herbal, Herbivore
--HEREO	HEHReeo	to Cling, Stick	Adhere, Adhesive, Cohere, Cohesion
HERPE--	HERpeh	Reptile, Snake	Herpetology, Herpetological, Herpetologist
HESPER--	HESSpur	West, Evening Star	Vespers, Hesperidium, *Hesperormis*
HETERO--	HETUR-oh	Different, Other	Heterosexual, Heterodox, Heterogeny
HEXO, HEXA--	HEKS-o, HAKSah	Six	Hexapod, Hex sign, Hexagon
HIBERN--	HIGHburn	Winter	Hibernaculum, Hibernal, Hibernate
--HIBIT	HIBit	Not Have, Not Hold	Inhibit, Prohibit, Inhibition, Prohibition
HIERO--	HIGHrō	Sacred, Secret	Hierograph, Hierocracy, Hierogram

ROOT/PREFIX/SUFFIX	PRONUNCIATION	MEANING	EXAMPLES OF USAGE
HIPPO	HIP-oh	Horse	Eohippus, Hippopotamus, Hippophile
HISTO--	HISST-oh	Tissue, Web	Histology, Histoid, Histogram
HOLO--	HO-LO	All, Entire	Holocaust, Holoblast, Hologamous
HOMEO--	HOMEee-oh	Sameness	Homeopath, Homeostasis, Homeotypic
HOMI--	HOME-ih, HAHM-ih	Mankind	Homicide, Human, Hominoid, Hominidae
HOMO--	HO-MO	Same	Homozygous, Homonym, Homosexual, Homoptera
HORA, HORO--	HŌrah, HŌrō	Hour	Horary, Horologe, Horological, Horoscope
HORREO--	HORee-oh	Bristle, Horrible	Horripilation, Horror, Abhorent
HUMO, HUMU--	HYU-mo, HYUMyoo	To Be Moist, Ground	Humus, Humiliate, Humble, Humic, Humic acid
HYALO--	high-AL-oh	Glass	Hyalogen, Hyaline membrane, Hyaloid
HYDRO--	HIDE-ro	Water	Hydraulic, Hydrant, Dehydrate, Hydrocarbon
HYETO--	highYET-oh	Rain	Hyetograph, Hyetology, Hyetographical
HYGRO--	HĪJ-rō	Wet	Hygrometer, Hygroscopic, Hygrophilous
HYLO--	HIGH-lō	Wood	Hylic, Hylozoic, Hylotheism, Hylicism
HYMEN	HIGHmen	Skin Membrane	Hymeneal, Hymenoptera, Hymenotomy
HYPER--	HIGHper	Above	Hyperactive, Hypercritical, Hypertonic
HYPHA--	HIGHfah	Weaving, Threads	Hyphaene, Hyphal, Hyphae, Hyphantria

ROOT/PREFIX/SUFFIX	PRONUNCIATION	MEANING	EXAMPLES OF USAGE
HYPNO--	HIP-no	Sleep	Hypnotherapy, Hypnotize, Hypnotic
HYPO--	HIPE-oh	Under	Hypodermic, Hypochondria, Hypotonic
HYPSO--	HIP-so	on High	Hypsography, Hypsometrical, Hypsophylum
HYSTERO--	HISStah-rō	Uterus, Womb	Hysteria, Hysterectomy, Hysteroid
--IASIS	EYEah-sis	Diseased condition	Myiasis, Psoriasis, Elephantiasis
ICHNEU--	ikNEW	Hunt	Ichneumon, Ichneumon fly, Ichnograph
ICON--	EYEkahn	Image	Iconic, Iconography, Iconoclast
ICT--	Ikt	Yellow, to Strike	Icteric, Ictus, Icterus
ICTHY--	IKthee	Fish	Icthyology, Icthyoid, Icthyosaur
--IDAE	ihDEE	Family	Ephemeridae, Culicidae, Hominidae
IDEA--	eye-DEAH	Form, See	Ideal, Ideate, Ideation
IDEO--	Ideo (as in "video")	Idea	Ideogram, Ideologist, Ideophone, Ideology
IDIO--	Idio (as in "video")	One's Own	Idiosyncracy, Idiom, Idiomatic, Idiopathy
IGNIS--	IGnis (as in "bigness")	Fire	Igneous rock, Ignite, Ignescent, Ignition
ILEO--	ILLee-o	Twist	Ileostomy, Ileitis, Ileum, Ileocolitis
IMAG--	IMMage, IMMag	Likeness, Image	Imagine, Image, Imago
IMBRIC--	IMbrik	Rain, Gutter Tile	Imbricate, Imbrex, Imbrication
IMID--	IMMid, immEYED	derived from Ammonia	Imide, Imidigen, Imido

ROOT/PREFIX/SUFFIX	PRONUNCIATION	MEANING	EXAMPLES OF USAGE
IMMUN--	imMYOON	Exempt	Immune, Immunity, Immunized, A.I.D.S.
IMPEDE, IMPEDI	imPEED, imPEDDee	Slow the Foot	Impede, Impediment, Impedance, Impedimenta
IMPET	IMPeht, imPETT	Rush Upon	Impetigo, Impetuous, Impetus
IN--	Inn	No, Not	Inability, Incoherent, Intangible
--IN	In	Pertaining to	Melanin, Epinephrine, Insulin
--INAE	ihNEE	Subfamily	Felininae, Meliponinae, Acrocerinae
INCENDI--	inSENDih	to Kindle	Incendiary, Incense, Incinerate
INCID--	INNsihd	Fall (in or upon)	Incident, Incidence, Angle of Incidence
INCIS--	INNsis, inSAISS	Cut into	Incisor, Incisive, Incisory
INDEX--	INNdeks	Point, Indicate	Index finger, Indexes, Index of Refraction
INDIC--	INNdick	to Declare	Indicate, Indicative, Indict
INDO--	INdough	Blue, Violet	Indigo, Indol, Indophenol
INFER--	inFEHR, inFUR	Lower, Later	Inferior, Infernal, Inferiority
INFLU--	INflew, INfluh	to Flow In	Influx, Influence, Influenza, Influential
INFRA--	INfrah	Under, Beneath	Infrastructure, Infrared, Infraclusion
INQUE--	INKweh	to Seek, Inquire	Inquest, Inquire, Inquisition, Inquisitive
INSULA--	INNsu-lah	Island	Insulate, Insulin, Peninsula
INTER--	INtur	Between, Inside	Interlude, Internal, Intervention
INTRA--	INtrah	Within	Intracerebral, Intramural, Intravascular

ROOT/PREFIX/SUFFIX	PRONUNCIATION	MEANING	EXAMPLES OF USAGE
--ION--	EYEahn	Violet Color	Anion, Ionize, Iodine, Iolite, Cation
IRIS--	EYE-riss	Rainbow Color	Iris, Iritis, Irisopsia, Iritomy
ISHIO--	ISSshe-o	Hip	Ischiectomy, Ischium, Ischiocele
ISO--	EYE-so	Equal	Isometrics, Isobar, Isosceles triangle
--ITIS	EYEtiss	Inflammation	Bronchitis, Gastritis, Neuritis
--JACT--	Jacked	Throw, Hurl	Jactitation, Ejaculate, Project, Eject
JUDIC--	JYOODik, JYOOdiss	Judge	Judicial, Judgment, Injudicious, Judicature
JUNCT	Junked	Join	Adjunct, Conjunction, Disjunction, Junction
--JUVEN--	JOOVen	Youthful	Rejuvenate, Juvenile, Juvenescent
KARY--	Carry	Nucleus	Karyokinesis, Karyolymph, Karyoplasm
KENO--	KEEN-oh	Emptying	Kenogenesis, Kenosis, Kenotron
KERET--	CAREut	Horny	Keratode, Keratoid, Keratoma
KILO--	KEY-low, KILLoh	Thousand	Kilometer, Kilogram, Kilowatt
KINE--	KINah, KEENeh	Moving	Kinetoscope, Kinetic energy, Kinetogenesis
LAB--	Lab, Layb	Labor, Change	Labile, Laboratory, Lability
LABIO--	LAHbee-o, LAYbee-o	Lip	Labia minora, Labial, Labialize
LABYRINTH	LABEHrinth	Curving Lane	Labyrinthine, Labyrinthic
LACCO--	LACKoh	Cistern	Laccolith, Laccolite, Laccolitic
LACE	LAYss	to Entice	Lace, Lacewings, Lacing

ROOT/PREFIX/SUFFIX	PRONUNCIATION	MEANING	EXAMPLES OF USAGE
LACER--	LASSsur	Mangled, Jagged	Lacerate, Lacerations, Lacerated
LACIN	LASSin	Flap	Laciniate, Lacinose, Laciniform
LACRIM--	LACKrim	Cry, Tears	Lacrimate, Lacrimator, Lachrymose
LACTO--	LACK-toe	Milk	Lactobacillus, Lactate, Lactose
LACU--	LAKyoo, LAHkoo	Pit, Basin	Lacuna, Lacunose, Lacustrine, Lake
LAMBDA	LAMBda	One who mis-pronounces the letter "L" (Lambda)	Lambdacism, Lambdoid, Lambda particle
LAMBEN--	LAMben	Lick Softly	Lambency, Lambently, Lambentness, Lamprey
LAMELA--	lahMELLah	Split into Thin Layers	Lamellated, Lamellar, Lamelliform
LAMELLI--	lahMELLIH	a Plate	Lamellibranch, Lamellose, Lamellicorn
LAMENT	laMENT	Bewail	Lamented, Lamentation, Lamentable
LAMIN--	LAMin	Thin Sheet, Scale	Laminate, Laminectomy, Laminitis
LAMPA	LAMPah	Shine	Lamp, Lampion, Lampad, Lampara
LANA--	LAHna	Wool	Lanate, Lanolin, Lanary
LANCE	LANss	Spear, Dart	Lance, Lancelet, Lancinate, Lancet
LANGUI--	LANguih	Faint, Weak	Languid, Languish, Languor
LANI--	LANih	Rend or Tear	Laniary, Lanner, Lanneret
LANUG--	LANyoog	Downy Wool	Lanuginous, Lanugo
LAPID--	LAPid	Stone	Dilapidated, Lapidary, Lapillus
LARD	Lard	Pig Fat	Lardaceous, Lardacein, Larder

ROOT/PREFIX/SUFFIX	PRONUNCIATION	MEANING	EXAMPLES OF USAGE
LARI, LARO--	LARih, LAIRoh	Ravenous Seabirds	Laridae, Larinae, Larine
LARVA	LARvah	Ghost (or "spirit" of)	Larviparous, Larvae, Larvate, Larval
LARYNG--	LAIRinj	Throat, Gullet	Laryngitis, Laryngoscope, Larynx
LATEN	LAYten	Lurking, Hidden	Latent, Latescent, Latency
LATER--	LATtur (as in "ladder")	Brick, Red	Laterite, Lateritic, Lateritious
LATI--	LATih	Broad	Laticlave, Latifoliate, Latitude
--LATRY, OLOTRY	LATrih, AHLAHtree	Extravagant Worship	Idolatry, Iconolatry, Litholatry
LATTIS	LATtiss (as in "gratis")	Lath	Space lattice, Lattice-work, Laticcing
LAUR--	Laural (as in "floral") or LA-oor (a dipthong, in which *both* vowels are sounded)	Laurel	Lauracae, Laureate, Laureled, Laurus
LAVO--	LAHVoh	Wash	Laundry, Lavabo, Lavage, Lave, Latrine
LAX	Lacks	Loose	Laxation, Laxative, Laxity, Lazy
LEGA--	LEGah	Bequeath	Legacy, Legal, Legatee, Legalization
LEGIS--	LEHjiss	Law-bearing	Legislate, Legislative, Legislator
LEGIT--	lehJIT	Lawful	Legitimate, Legitimist, Illegitimate
LEGO--	LEJoh	to Read	Lector, Lecture, Lectern
--LEGO	LEGoh	Select, Choose	Elect, Legation Delegate, Legible
LENS	Lenz	Lentil	Lenticel, Lenticular, Lentils, Lentoid
LEO	LEEoh, LEHoh	Lion	Leonine, Leopard, Leonid

ROOT/PREFIX/SUFFIX	PRONUNCIATION	MEANING	EXAMPLES OF USAGE
LEPIDO	LEPih-dough	Scale, Flake	Lepidoptera, Lepidote, Leprosy
LEPTO--	LEPToh	Peel	Leptome, Leptomatic, Leptosporangiate
LEPUS	LEEPus	Hare	Leveret, Leperine, Leperinae
LEVATA	lehVAHTah	to Raise (up)	Elevator, Levant, Levee, Levator
LEVATA	lehVAHTah	to Raise (money)	Levy, Levies, Leviable
LEVE--	LEVeh	Balance	Level, Leveling, Lever
LEVE, LEVO--	LEHvo or LEEVoh	to the Left	Levogyrate, Levorotatory, Levulose
LEX--	Leks (rhymes with "decks")	Read, Words	Lexicon, Dyslexia, Lexicology, Legend
LIBRA--	LEEBrah	Balance	Librate, Libratory, Libration of the moon
LIBRA, LIBRO--	LEEBroe, LAIBrah	Belonging to Books	Libel, Library, Libretto
LIBRI--	LIEbrih	Tree Bark	Libriform, Liber, Liberaceous
LEICHO--	LIKEoh	Lick	Lichen, Lichenaceous, Lichenologist
LIGA--	LEEgah or LIGGah	to Bind	Ligature, Ligament, Ligate, Liable
LIGNI--	LIGnee	Wood	Lignin, Lignose, Lignite, Ligneous
LIGULA	LIG-YOOlah	Little Tongue	Ligular, Ligulate, Ligule
LIMBUS	LIMbus	Edge	Limbate, Limb, Limbic, Limbo
LIMIN, LIMIL	LIHmin	Threshold	Limen, Liminal, Eliminate, Limit
LIMNO--	LIHMno (as in "limb-know")	Bodies of Water	Limnology, Limnic, Limnetic
LINEA--	LINih-ah	Line	Lineage, Lineal, Lineolate

ROOT/PREFIX/SUFFIX	PRONUNCIATION	MEANING	EXAMPLES OF USAGE
LINGUA--	LEENgwah	Like a Tongue	Lingulate, Linguistics, Lingual
LINUM	LINum	Flax	Linoleum, Linnet, Linen, Lingerie
LIPE, LIPO--	LIPeh, LIP-oh	a Fat	Lipase, Lipoid, Lipoma, Lipemia
LIQUA	LICKwah	Be Fluid	Liquate, Liquescent, Liquor, Liquid
LITERA	LITERah	Letter	Literary, Literature, Illiterate
--LITH--	Lith	Stone, Stonelike	Lithograph, Monolith, Neolithic
LITIGA--	LITTih-gah	to Strive	Litigation, Litigate, Litigant
LITRA, LIBRA	LEEtrah, LEEBrah (The r's should be trilled.)	Pound, Weight	Lira, Liter, Lb.
LIXIV--	LICKsiv	Lye, Ashes, Alkali	Lixivial, Lixivium, Lixiviation
LOCO--	LOW-ko	Place, in Place	Locus, Location, Locomotive
LOCUT--	LOWkut	to Speak	Eloquent, Locutory, Elocution, Loquacious
LODIC--	LAHdik	Little Blanket	Lodicule, Lodicle, Lodicula
--LOGO--	LOW-go	Speech, Thought	Logical, Dialogue, Eulogy, Apology
--LOGY	LOW-jee	Study	Arachnology, Physiology, Meteorology
LONGI--	LAWNjih	Long	Longitude, Longicaudal, Longicorn
LOPHO--	LOAF-oh	Crest	Lophodont, Lophobranch, Lopholith
LORICA	loRIKE-ah	Thong (to secure plates of armor)	Loricata, Loricate, Loricati
LUBRI--	LOOBrih	Slippery	Lubricious, Lubricant, Lubricate

ROOT/PREFIX/SUFFIX	PRONUNCIATION	MEANING	EXAMPLES OF USAGE
LUCI--	LOOSih	Light	Lucid, Luciferin, Luciferase, Lucifer
LUCR--	LOO-kr	Gain	Lucre, Lucrative, Lucratively
LUCUBRA	looKOOBrah	Light, Night-study	Lucubrate, Lucubration, Lucubratory
--LUD--	Lood	Play	Elude, Allude, Elusive, Ludicrous
LUMBU--	LOOMbuh	Loin	Lumbar, Loin cloth, Lumbago
LUMEN--	LOOmen	Light	Illuminate, Luminescent, Lumenophore
LUNA--	LOONah	Moon	Lunatic, Lunar, Lunate, Lunacy
LUNI--	LOONih	Moon-like	Luniform, Lunisolar, Lunitidal
LUNUL--	LOONyool	Crescent	Lunula, Lunulated, Lunular
LUPUS--	LOOPus	Wolf	Lupine, Lupus vulgaris, Lupoma
LUSTR--	LUSTr	Shine	Luster, Lustrum, Lustrin, Illustrate
LUTEO	LOOTih-oh	Golden Yellow	Luteolin, Corpus luteum, Luteous
LUXU--	LUCKS-oo	Extravagance	Luxuriant, Luxuriate, Luxury
LYCO--	LIKE-oh	Wolf	Lycanthrope, Lycopod, Lycomorph
LYMPH--	LIMPf	Clear Fluid	Lymphocyte, Lymphoid, Lymphoma
LYO--	LYE-oh	Loose	Lyophilic, Lyophobic, Lyogenic
LYRI--	LEERih	Harp, Music	Lyrical, Lyrics, Lyricism, Lyre
--LYSIS--	LYE-sis	Loosen, Break	Hydrolyze, Analyze, Dialysis, Lysosome

ROOT/PREFIX/SUFFIX	PRONUNCIATION	MEANING	EXAMPLES OF USAGE
MACRO--	MACK-roe	Large, Long	Macropsia, Macrophage, Macroscopic
MACU--	MACKyoo	Spot	Mackerel, Mackle, Macula, Immaculate
MADRE--	MAHDray (The r should be trilled.)	Mother	Madreporaria, Medreporic, Madreporite
MAGIC	MAJik	of the Magi, Magical	Magician, Magic Lantern, Magical
MAGIST--	MAJist	Master	Magisterial, Magistrate, Magistrature
MAGMA	MAGma	to Knead Dough	Magma, Magmatic, Massecuite
MAGNA--	MAGna	Great	Magnate, Magnum, Magna Carta
MAGNES--	MAGnes	an Area in Greece; a Substance Containing Magnesium	Magnesia, Magnesite, Magnesium light
MAGNET--	MAGnet	Expressing Strong Attraction	Magneto, Magnetic, Magnetize, Magnetite
MAGNI--	MAGnih	Large	Magnify, Magnificent, Magnitude
MAL--	Mal (as in "pal")	Faulty	Malodorous, Malocclusion, Malpractice
MALACOS	MALLAHkoss	Soft	Malachite, Malacology, Malacophilous, Malaxate
MALE--	MAHla or MALLeh	Evil	Malediction, Malefactor, Malevolent
MALI--	MALLih	Bad	Malicious, Malign, Malignant
MALL--	Mall (as in "pal")	Hammer	Malleable, Malleate, Malleolus, Mallet
MALLO	MALLoh	Soft	Mallow, Rose mallow, Marshmallow

ROOT/PREFIX/SUFFIX	PRONUNCIATION	MEANING	EXAMPLES OF USAGE
MALUM	MALLum	Apple	Malic acid, Malaceous, Malate
MANDI--	MANdih	Jaw, Chew	Mandible, Mandibular, Mandibulary
MANI--	Many	Hand	Manipulate, Manicure, Maniple
MANI--	MANih	Struck by the Hand	Manifest, Maniple, Manifold
MANIA--	MAYnee-ah	Compulsion	Maniac, Manic, Manic-depressive
MANU--	MANyoo	a Hand	Manufacture, Manumission, Manual
MAR--	Mar	Sea, Ocean	Marine, Maritime, Meremma, Marinate
MARGAR--	MARJar or MARGar	Pearl, Oyster	Margarine, Margarite, Margaric acid
MARGI--	Margie	Border	Margin, Marginalia, Marginal
MASTO--	MAST-oh	Breast, Meat	Mastodon, Mastoid, Mastopathy
MATERI--	maTEARY	Matter	Material, Materialistic, Materialize
MATH--	Math	Learn	Mathematics, Mathematician, Mathematical
MATIN	MAHtin	Morning	Matins, Matineé, Matinal, Matutinal
MATR--	MAHtr	Mother	Matricide, Matriarchy, Matrilocal
MATRICULA	maTRICKyoo-la	Enrolled, Public Register	Matriculate, Matriculant, Matriculation, Matrix
MATUR--	MAtyoor or maTYOOR	Ripe	Maturation, Mature, Maturity, Immature
MAXI--	MAXih	Great, Highest	Maxim, Maximum, Maximal
MAXILL--	MAXil	Upper Jaw	Maxilla, Maxilliped, Maxillary

ROOT/PREFIX/SUFFIX	PRONUNCIATION	MEANING	EXAMPLES OF USAGE
MAXIM	MAXim	Great	Maximize, Maximalist, Maximization
--MEA, MEO	MEah, MEoh	to Pass (through)	Permeate, Permeability, Permeation
MEANDER	meANDer	Winding Path	Meandering, Meandrous, Meander
MECHANI--	meh-KAHN-ih	Machine	Mechanical, Mechanic, Mechanize
MEDAL	MEDDal	Metal	Medalic, Medalist, Medal of merit
MEDDL--	MEDDl	Mix	Meddler, Miscellaneous
MEDI--	MEDDih	Middle	Mediterranean, Medieval, Mediocre
MEDIA--	MEEDYah	Middle	Median, Mediate, Mediastinum
MEDIC--	MEDik	Physician, Heal	Medicine, Medical, Medication
MEDIT--	MEDit	to Consider	Meditate, Meditation, Meditative
MEDULLA	mehDOOlah	Marrow	Medullary, Medullated, Medulla oblongata
MEI--	My	Smaller (in size or in number)	Meiogenic, Meiosis, Meiotic
MEGA--	MEGah	Great	Megaphone, Megacephalic, Megalith
MEGALO--	MEGAH-lo	Million (times)	Megalomania, Megalocephalus, Megalosaur
MELAN--	MELLahn	Black	Melancholy, Melanin, Melanoma
MELANO--	MELLah-no	Looking Black	Melanoid, Melanosis, Melanotic, Melanochroic
MELI--	MELLih	Honey	Melilot, Mellifluent, Melinite, Melituria
MELIOR--	MEELyor or MELLyor	Better, Amend	Ameliorate, Ameliorated, Meliorism

ROOT/PREFIX/SUFFIX	PRONUNCIATION	MEANING	EXAMPLES OF USAGE
MELOD--	Mellowed	Song	Melodeon, Melody, Melodramatic
MEMBRANE	memBRAIN	Thin Parchment	Membranous, Membraniferous, Tympanic membrane
MEMOR--	MEMMor	Mindful	Memorandum, Memoir, Memorial
MENAC--	MENNus	Threats	Menace, Menacing, Menaced
MENDAC--	menDĀSE	Lying, False	Mendacious, Mendacity, Mendaciousness
MENDICO	MENDIH-ko	Beg	Mendicant, Mendicancy, Mendicity
MENE, MENO--	MENeh, MEN-no	Moon	Meniscus, Menopause, Menstruation
MENING--	menINj	Membrane	Meninges, Meningeal, Meningitis
MENSA--	MENsah	Table, Food	Mensal, Commensalism, Commensal
MENSU--	MENsu	to Measure	Mensurable, Mensuration, Mensurative
MENTA--	MENtah	Mind	Mental, Mentality, Mention, Demented
MENTAL	MENtal	the Chin	Mental point, Mentonniere, Mental, Mentum
MENTH--	Menth	Mint	Mentholaceous, Menthol, Menthene
MEPHI--	MEFFih	Stinking Exhalation	Mephitic, Mephitis, Mephitism
--MER	Mehr	Form, Shape	Polymer, Blastomere, Metamerism
MERCENA--	MERSENah	Reward, Trade	Mercenary, Merchant, Mercer, Mercy
--MERE, MERO	Mehr, MEHRo	Part	Blastomere, Merognostic, Metamere

ROOT/PREFIX/SUFFIX	PRONUNCIATION	MEANING	EXAMPLES OF USAGE
MERGO	MURgo	Dip	Immerse, Submerse, Merganser, Merge
MERI--	Merih	Divide	Meristem, Meridian, Meristic
MESENCHYMA	mehSINN-KEYmah	Middle-juice	Mesenchymal, Mesenchymatic, Mesenchymatous
MESI--	MESZih	Middle	Mesial, Mesian, Mesially, Mesial plane
MESO--	MESZoh	Middle	Mesoblast, Mesoderm, Mesozoic, Mesonotum
META--	METTah	Among, Changing	Metaphase, Metastasis, Metatarsal
METALL--	METal	Metallic Substances	Metalloid, Metallography, Metallurgy
METEOR	MEATY-or	Raised Above	Meteorgraph, Meteor, Meteorology
METH--	Meth	Wood	Methane, Methyl, Methule, Methylene
--METR--	METTr	Measure	Trigonometry, Metronome, Centimeter
METRO	METro	Mother, Womb	Metropolis, Metritis, Metronymic
MEZZO--	METTzoh	Half, Medium	Mezzanine, Mezzo-Soprano, Mezzotint
MIASM--	myAZM	Polluting Exhalations	Miasma, Miasmatic, Miasmic
MICRO--	MY-kro	Small	Microbe, Microscope, Micrometer
MIGRO--	MY-grō	Remove, Leave	Migrate, Migrant, Immigrant, Emigrate
MILIAR--	MILLiar	Millet	Miliaria, Miliary, Milium, Milomaize
MILIT--	MILL-it	Soldier	Military, Militant, Militia
MILLI--	MILLih	Thousand(th)	Millenium, Mile, Milliampere, Millipede

ROOT/PREFIX/SUFFIX	PRONUNCIATION	MEANING	EXAMPLES OF USAGE
MIME--	Myme	Mimic, Copy	Mimosa, Mimotype, Mimeograph
MINE	Mine	Burrow, Mine	Mineral, Miner, Minethrower, Mineralization
MINI--	Minnie	Least, Small	Minimal, Miniature, Minimum
MINIS--	MINNiss	Assistant, Help	Minister, Administer, Ministry
MINU--	MINNyoo	Less	Minute, Minuend, Minuet, Minus
MIO--	MY-oh	to Shut, Less	Miotic, Miocene, Miosis, Myopia
MIRA, MIRU--	MIRah, MIRuh	Wonderful	Miraculous, Miracle, Mirage
MISC--	Miss	Mix	Miscible, Miscellaneous, Miscellany
MISER	MISSer, ME-sir	Wretched	Misery, Miserly, Miser, Miserable
MISO--	MEEso	Hate	Misogyny, Misanthrope, Misogamy
MISU, MITTO	MISSu, MITTo	Send	Missile, Message, Remit, Remission, Mission
MITIS	MĪTE-iss or MITTiss	Mild	Mitigate, Mitis casting, Mitigation
MITO--	MĪTE-oh	Thread	Mitochondria, Mitosis, Mitotic
--MOBIL--	MO-bill	Movable	Automobile, Immobility, Mobility, Mobilize
MODUS	Mahduss, Mōduss	Measure, Regulate	Model, Mode, Moderate, Modality, Module
MOLA--	MOlah	Mill, Grind	Molar, Molaris, Molary, Mole
MOLE--	Mole	Mass, Weight	Molecule, Molecular, Molest, Mole

ROOT/PREFIX/SUFFIX	PRONUNCIATION	MEANING	EXAMPLES OF USAGE
MOLLUSC	MAHLusk, MŌ-lusk	Soft	Mollusca, Mollify, Mollusk
MOLYBD--	moLIBBid	Lead	Molybdenite, Molybdenous, Molybdic acid
MOMENT	MOment	Movement, Balance	Momentum, Momentarily, Momentous
MONA	MŌnah	Unit, Single	Monad, Monadic, Monanthous, Monandrous
MONETA	mōNETTah	Coin, Mint	Money, Monetary, Monetize
MONI--	MAHnih, MOHneh	Warn	Admonition, Monitor, Monition, Admonish
MONO--	MO-no	One, Alone	Monotone, Monarch, Monochrome
MONUMENT	MAHN-YOOment	Remind	Monument, Monumental, Monumentalize
--MONY	MUNnee, MOANnee	Money	Alimony, Parsimony, Money
MORB--	Morb	Disease	Morbidity, Moribund, Morbid
MORDA--	MORdah	Bite	Mordaceus, Mordant, Morsel
MORMO--	MORmo	Black	Moor, Morel, Morion
--MORPH--	Morf	Form	Morphology, Polymorphic, Mesomorph
MORT--	Mort	Dead, Death	Mortify, Mortuary, Mortal
MORU--	MORoo	Mulberry	Morula, Morin, Morion
MOT--	Moat	Motion	Motile, Motion, Motive, Motor
MOV--	Move	Move	Movement, Movable, Movies
MUC--	MYOOss, Myook	be Moldy	Mucilage, Mucedinous, Mucin, Mucus

ROOT/PREFIX/SUFFIX	PRONUNCIATION	MEANING	EXAMPLES OF USAGE
MULS--	MULss	Milk	Emulsion, Emulsoid, Mulct, Emulsify
MULTI--	MULTih	Many	Multiple, Multiparous, Multipolar
MUNICIP--	myooNISSip	Duty-take	Municipal, Municipality, Municipalize
MUNIO	MYOON-yo	Fort, Fortify	Muniment, Munition, Ammunition
MURAL	MYOORal	Wall	Murals, Muriform, Murine, Intramural
MUSCA	MOOSkah	a Fly	Muscarine, Muscid, Mosquito, Muscidae, Musket
MUSCL--	MUSSul	Muscle	Muscled, Muscle, Musculature, Myalgia
MUTA--	MYOOtah or MOOtah	Change	Mutation, Mutagen, Mutate, Mutual
MYCE--	Mice	Nail, Wart	Mycelium, Myceloid, Mycelia
MYCO--	Mike-o	Fungus	Mycobacterium, Mycosis, Mycostat
MYDRI--	My-Dry	Pupil Dilation	Mydriasis, Mydrine, Mydriatic
MYELO--	myELLo	Marrow, Spinal Cord	Encephalomyelitis, Myelocyte, Myeloid
MYO--	MY-oh	Muscle	Myocardium, Myoclonus, Myopathy
MYOP--	MY-ōp	Eye-close	Myopia, Myopic, Myope
MYRI--	MEERih	Ten-thousand	Myriagram, Myriad, Myriapod
MYRME--	MURMeh	Ant	Myrmecology, Myrmecophilism, Myrmex
MYRO--	MIRR-oh, MY-ro	Ointment	Myrobalan, Myrosin, Myrrh

ROOT/PREFIX/SUFFIX	PRONUNCIATION	MEANING	EXAMPLES OF USAGE
MYSTE--	MISSteh	One Initiated; Close to the Lips	Mystery, Mysterious, Mystical
MYXA--	MIXah	Slime, Mucus	Myxamoeba, Myxinidae, Myxomycete
NARCO--	NARK-oh	Numbness, Torpor	Narcolepsy, Narcotic, Narcosis
NAS--	NAYss or Nass	Nose	Nasiform, Nasology, Nares, Nasopharyngeal
--NASCENT--	NAYsunt	Arising	Neonascent, Nascency, Nascent
NATA--	NAYtah	Born	Neonatal, Nation, Native, Nativity
NATA--	NAHtah or NATah	to Swim	Natant, Natation, Natatorium
NAUS--	NAH-ooss (as in "mouse")	Ship	Naumachia, Nausea, Nauseated
--NAUTIC--	NAWtik	Sailor	Nautical, Astronaut, Cosmonaut, Nautilus
NAVI--	NAHVih	Ship	Navy, Navigate, Navicert, Navicular
NEBUL--	NEBByool	Fog, Mist	Nebula, Nebulous, Nebulize
NECRO--	NECKcrow	Corpse, Dead	Necrotic, Necrophilia, Necropolis
NEGAT--	NEGGaht	to Deny, Opposite	Negative, Negate, Negation, Negativism
NEKTO--	NECKED-oh	Swim (near the sea's surface)	Nekton, Nekteric, Nektonic
NEMATO--	nehMATT-oh	a Thread	Nematocyst, Nematode, Nematoform
NEO--	NEE-oh	New, Recent	Neonatal, Neolithic, Neocene, Neon
NEPHO--	NEFFoh	Cloud	Nephoscope, Nephology, Nephogram

ROOT/PREFIX/SUFFIX	PRONUNCIATION	MEANING	EXAMPLES OF USAGE
NEPHRO--	NEFFrō	Kidney	Nephritis, Nephrocele, Nephrotomy
NERI--	Nerih	Sea Nymph, Fish	Nereid, Neritic zone, Nereus
NEURO--	NEWrō	Nerve, Sinew	Nervous, Neuron, Neuroma, Neuroblastoma
NEUTRO--	NEWTrō	Neutral	Neutron, Neutrosphere, Neutrino
NICTI--	Nikti	Wink, Beckon	Nictitate, Nictitation, Nictitating membrane
NIDU--	NIDyoo	Nest, Center	Nidus, Nidulant, Nidology
NITR--	NIGHTer	Soda	Nitrate, Nitrogen, Niter
NOCTI--	KNOCKtih	Night	Nocturnal, Noctilucent, Noctograph, Nocturn
NODO, NODE	NO dough, Nōde	Knot, Swelling	Nodule, Nodular, Nodulous, Nodical
--NOM	Nome	Law, Govern	Autonomy, Astronomy, Economy, Taxonomy
NOMA, NEMA	NOmah, KNEEmah	Roving, Grazing	Noma, Nomad, Nomadic, Nomadism
NOMEN	KNOWmen	Name	Nomenclature, Nominal, Nominate
NON--	Non (as in "John")	Not	Nonattendance, Nonessential, Nondescript
NONA--	NAHna	Nine	Nonagenarian, Nonane, Nonagon, November
NORMA	Normah	Rule	Normal, Normality, Abnormal, Normalcy
NOSO--	NO-so	Disease	Nosogeny, Nosomania, Nosology
NOTA--	NOTE-ah	Know, Notable	Notary, Notation, Notarize, Note, Notion
NOTO--	NO-toe	the Back	Notochord, Notoderm, Notomorph

ROOT/PREFIX/SUFFIX	PRONUNCIATION	MEANING	EXAMPLES OF USAGE
--NOVA--	NOvah	New	Novacaine, Novelty, Supernova, Innovation
NOXA	KNOCKSah	Hurt, Harm	Obnoxious, Noxious, Noxal, Noxiousness
NUCLE--	NEWKleh or NOOKleh	Nut, Walnut	Nuclear, Nucleoplasm, Nucleolus
NULL--	Null	None, No	Nullify, Nulliparous, Nullipore, Nullification
NUTRI--	NEWTrih	Nurse, Feed	Nourish, Nourishment, Nourishable, Nutrient
NYCTA--	NICKtah	Night	Nyctalopia, Nyctitropism, Nyctalopic
--NYM	Nimm	Name	Pseudonym, Homonym, Anonymous
NYMPH--	Nimf	Bride	Nymphalid, Nymphomania, Nympholepsy
OB--	Ahhb	To, Facing	Observe, Obstinate, Obtrude
OBSTETRI--	ahbSTETTrih	Before-stand, Midwife	Obstetrical, Obstetrics, Obstetrician
OBVER--	AHBvur	Toward, Turn	Obvert, Obverse, Obversion, Obvolute
OCCIPIT	oxSIPPit	Back of Skull	Occipital, Occipital lobe, Occiput
OCCLU--	OHkloo or AHKloo	to Close, Shut	Occlude, Occlusion, Occlusal
OCCULT	oh-KULT	Hidden	Occultation, Occultis, Occultism
OCHRE	OAKur	to Dye (with red or yellow)	Ochre, Ochroid, Ocherous
OCTO--	AHKToh	Eight	October, Octagon, Octopus, Octave
OCULUS	AHKyool-us	Eye	Monocle, Binoculars, Monocular, Oculist

ROOT/PREFIX/SUFFIX	PRONUNCIATION	MEANING	EXAMPLES OF USAGE
--ODO--	ŌDE-oh	Way, Path	Odometer, Anode, Cathode
ODONT--	ohDON'T	Tooth	Odontalgia, Odontoblast, Odontology
ODOR--	ŌDE-ur	Stench, Odor	Odoriferous, Odorless, Odorous
OENO--	EEN-oh	Wine	Oenanthic, Enologist, Oenomania
OESTR	ESStr (sometimes WEstr)	Gadfly, "Heat"	Estrogen, Estrus Estriol
OFFEN--	ohFEN	Before-strike	Offensive, Offend, Offensively
OFFIC--	OAFiss or AHFiss	Work-do	Office, Officer, Official, Officious
--OID	Oyd	Like, Form	Hominoid, Ovoid, Spheroid, Mongoloid
--OL	OHl or All	Alcohol, Oil	Methanol, Glycol, Ethanol, Isopropenol
OLEO--	OLEE-oh	Oil	Oleomargarine, Oleoginous, Oleate
OLFACTOR--	ol-FAKTor	Smelling-bottle	Olfaction, Olfactometer, Olfactory
OLIGO--	OHlee-go	Few	Oligarchy, Oligochaete, Oligodontia
OLIVA	OHleev-ah	Olive	Olivary, Olive, Olivacious, Olivenite
--OLOGY	ALAHjee	Study of	Nosology, Entomology, Cytology
--OMA	OHMmah	Tumor	Hematoma, Carcinoma, Glaucoma
OMBRO--	OHMbro	Rainstorm	Ombrophyte, Ombrology, Ombrometer
OMMA--	OHMma	Eye	Ommateum, Ommatidium, Ommatophore
OMNI--	AHMnih or OHMnih	Many	Omnipotent, Omnibus, Omnivore

ROOT/PREFIX/SUFFIX	PRONUNCIATION	MEANING	EXAMPLES OF USAGE
OMPHALO	ohmFAHL-oh	Navel	Omphalocele, Omphalos, Omphaloskepsis
--ON	Ahn	The	Neutron, Neon, Nephron, Proton
ONEIRO--	oh-NIGHro	a Dream	Oneiric, Oneirology, Oneiromancy
ONOMA--	ONoh-ma	Name	Onomastic, Onomatomania, Onomatopoeia
ONTO--	AHNT-oh	Being	Ontogeny, Ontological, Ontology
OÖ	Oh-Oh	Egg	Oögenesis, Oöcyte, Oögonium, Oölith
OPERA	AHPERah	Work	Opus, Operand, Operate, Operator
OPERC--	ohPERK	a Cover	Operculum, Opercle, Opercular, Operculate
OPHI--	OAFih	Serpent	Ophidia, Ophidian, Ophiolatrous, Ophiology
OPHTHAL--	AHFthal	of the Eye	Ophthalmology, Ophthalmic, Ophthalmoscope
OPIN--	ohPINE	Think	Opinion, Opine, Opinionated
OPISTHO--	ohPIStho̅	Behind	Opisthenogenesis, Opisthognathous, Opisthographic
OPIUM	OPih-um	Juice	Opiumism, Opiate, Opodeldoc, Opium
OPPO--	ohPO	Before	Opponent, Opportune, Oppose, Opposite
OPPRES--	ohPRES	Press Against	Oppress, Oppression, Oppressive
OPPROB--	ahPRŌBE	Disgrace Upon	Opprobrium, Opprobrious, Opprobriously

ROOT/PREFIX/SUFFIX	PRONUNCIATION	MEANING	EXAMPLES OF USAGE
--OPSIS--	AHPsis	Appearance	Synopsis, Macropsia, Opsiform
OPSON--	AHPson	Harvest, Food	Opsonification, Opsonic, Opsonize
OPT--	AHPT	Choice	Option, Optative, Opt, Optimal
OPTO, OPTI	AHP-toe, AHP-tih	to See	Optical, Optometrist, Optician
OPUGN, REPUGN--	ohPUGn, rePUGn	Fight Against	Oppugnance, Pugnacious, Repugnant, Impugn
ORA--	ORE-ah	Mouth, Pray	Oral, Oracle, Orifice, Orator
ORB--	Orb	Circle	Orbit, Orb, Orbiculate, Orbital
ORCHIDO--	ORKid-oh	Testicle	Orchidectomy, Orchidotomy, Orchitis
ORDINO	ORDih-no	Set in Order	Ordain, Ordinance, Order, Ordinal
ORGANO--	orGAN-oh	Organ (containing carbon)	Organophosphate, Organogenesis, Organotherapy
ORIG--	ohRIJJ	Rise, Beginning	Original, Originate, Originator
ORNITH--	ORNith	Bird	Ornithopter, Ornithic, Ornithology
ORNO	ORN-oh	to Adorn	Ornament, Ornate, Adornment, Ornamental
ORO--	OR-oh	Mountain	Orometric, Orology, Orographic cloud
ORTHO--	ORTH-oh	Straight	Orthodox, Orthodontist, Orthoptera
--ORY	Ory	Place	Dormitory, Factory, Rectory, Laboratory
OS--	AHSS	a Bone	Ossify, Ossification, Ossuary

ROOT/PREFIX/SUFFIX	PRONUNCIATION	MEANING	EXAMPLES OF USAGE
OSCILL--	AHSSil	to Swing	Oscillate, Oscillogram, Oscillations
OSCUL--	AHSKyool	Kiss, Mouth	Osculum, Osculation, Osculant, Oscule
--OSE	Owes	Fullness, Origin	Sucrose, Verbose, Dextrose, Bellicose
--OSIS	OHsis	Abnormal Condition	Neurosis, Tuberculosis, Varicosis
OSM	AHSm, Olizm	Odor, Smell	Osmic, Osmious, Osmium, Ozone
OSMOS--	AHZmos, OWESmos	Pushing	Osmose, Osmosis, Osmotic pressure
OSTE--	OWEStee or AHSTih	Bone	Osteal, Osteoblast, Osteomyelitis
OSTEN--	AHSStent	Exhibit, Reality	Ostensible, Ostensory, Ostentation
OSTI--	AHSTih	Mouth, Door	Ostiary, Ostiole, Ostiolar
OSTRA--	AHSTrah	Oyster, Shell	Ostracize, Ostracod, Ostracean
OTO, OTI--	OHtoe, OHtih	Ear	Otocyst, Otoscope, Otology, Otitis
--OUS	Us	a Quality	Precocious, Dexterous, Ambitious, Oviparous
OVI, OVO	OWEvih, OWE-vō	Egg	Ovipositor, Ovoid, Oviparous, Oval
OXALO, OXALI--	oxSALL-oh, ahkSALLY	Sharp, Acid	Oxalic, Oxalic acid, Oxalis
OXI, OXY	OXih, OXee	Rust, Acid	Oxidize, Oxygen, Oxyacetylene
OXYGON	OXIH-gahn	Triangle (with three acute angles)	Oxygonal, Oxygone, Oxygonial
OZO--	OWES-oh	Smell	Ozone, Ozostomia, Ozonosphere, Ozena
PAEDO--	PEED-oh or PAYED-oh	Child, Larva	Pedagogue, Pedokinesis, Pedophile

ROOT/PREFIX/SUFFIX	PRONUNCIATION	MEANING	EXAMPLES OF USAGE
PAGI--	PAHJee	Scaffold, Leaf	Pagina, Pagination, Pageant
PALEA	PAYlih-ah	Chaff, Soft	Palliasse, Pallet, Palea, Paleate
PALEO--	PAYlih-oh or PALih-oh	Old, Ancient	Paleology, Paleolithic, Paleocene
PALI--	PALih	Repeat, Again	Palingenesis, Palinode, Palisade
PALID	PALid	Pale	Paleness, Pallium, Pallid, Pallor
PALM	Pahlm	Hand, a Tree	Palmar, Palma, Palmaceous, Palmary
--PALP--	PALp	Lip	Palpus, Pedipalp, Protopalp
PALPA--	PALpah	Touch	Palpae, Palpable, Palpation
PALUD--	pahLOOD	a Swamp	Paludal, Paludamental, Paludism
PAN--	Pan	All	Panacea, Pandemic, Pan-American
PAPILL--	PAP'll	Nipple	Papillae, Papilloma, Papillary, Papescent
PAR--	Par	Bring Forth	Parent, Parturition, Parthenogenesis
PARA--	PAIRah	Beside	Parasite, Paramecium, Parabola
PARES--	pahREES	Beside, Let-go	Paresthesia, Paresthesis, Paresis
PARIET--	pahRITE	Wall	Parietal, Parietal bone, Paries
--PAROUS	PAIRus	Bearing Young	Viviparous, Nulliparous, Oviparous
PARTI--	PARTih	Part, Divide	Partisan, Partition, Partite
PARTICIP--	parTISSip	Take-part	Participate, Participant, Participle

ROOT/PREFIX/SUFFIX	PRONUNCIATION	MEANING	EXAMPLES OF USAGE
PASTI--	PASStih	Loaf of Bread	Pastille, Pantry, Panada, Pannier, Pastry
PATE--	PATeh	Open, Shallow Pan	Patent, Paten, Patella, Patera, Pateriform
PATER	PAHTTur	Father	Paternal, Paternity, Paterfamilias
--PATHO--	Path-oh or PAHtho	Suffer, Illness	Pathetic, Pathology, Pathogen, Psychopath
PATRI--	PAHTrih or PAYtrih	Father	Patriotic, Patrilinial, Patriarch, Patrician
PATUL--	PAT-chewul	Open, Wide Pan	Spatulate, Patulous, Patio, Paten
PECCO--	PECK-oh	Sin	Peccable, Impeccable, Peccant, Peccadillo
PECTI--	PECKTih	a Comb	Pectinate, Pectate, Pectinella
PECTIC	PECKtik	Make Solid (gel)	Pectic acid, Pectin, Pectose
PECTUS	PECKEDus	Breast	Pectoral, Pectoral arch, Pectoralis
PECULA--	PEKyoo-lah	Theft	Peculate, Peculation, Peculator
PECULI--	PECKYOO-lee	One's Own, Cattle	Peculiar, Peculiarity, Peculium, Pecuniary
PED--	PEHdd	Foot	Pedestrian, Pedal, Pedicel, Pedipalp
PEDA--	PEDdah	Slave Who Leads Child	Pedagogue, Pedagogic, Pedant, Pedantic
PEDERA--	PEDDERah	Boy-lover	Pederast, Pederasty, Pederastic
PEDI--	PEDih	Little-foot	Pedicle, Pedicellaria, Pedicellate
PEDO--	PEHD-oh or PEED-oh	Reproduction (by an immature animal)	Paedogenesis, Pedobaptist, Paedomorphic
PEDUNC--	pehDUNK	Stalk, Stem	Peduncle, Peduncular, Pedunculated

ROOT/PREFIX/SUFFIX	PRONUNCIATION	MEANING	EXAMPLES OF USAGE
PELAG	pehLAYj	the Open Sea	Pelagian, Pelagic, Pelasgian
PELL--	PELL	Skin, Pelt	Pelage, Pell, Pelt, Pellagra, Pellicle
PELLUC--	pehLOOSE	Shine through	Pellucidity, Lucid, Pellucid
PELOR--	pehLOR	Monster	Peloria, Pelorism, Peloriate, Peloric
PELTAS--	PELTus	Small Leather Shield	Peltast, Peltate, Peltry
PENA--	PEENah	Punishment	Penal, Penalize, Penalty, Penance
PEND--	Penned	Something Hanging	Pendant, Pendulous, Pendicle, Dependent
PENE, PENU--	PENeh, PENnoo	Almost	Peneplain, Peninsula, Penumbra, Penultimate
PENETR--	PENnetr	Put within	Penetrate, Impenetrable, Penetralia
PENICIL--	PENnih-sill	Painter's Brush	Pencil, Penciling, Penicillin
PENIT--	PENnit	Repent	Penitent, Penitentiary, Penance, Penology
PENNA	PENnah	Feather	Pennant, Pennated, Pennate, Pennon
PENTA--	PENTah	Five	Pentagon, Pentadactyl, Pentarchy, Pentane
PEPO	PEP-pō	Melon	Pepino, Peponida, Peponium
PEPSIS	PEPsis	Cooking, Digest	Pepsin, Dyspeptic, Peptogen, Peptone
PER--	Purr or Pehr	Through	Perjury, Permeable, Perspective
PERCEP--	perSEHP	to Take By	Perceive, Perception, Perceivably, Perceptive
PERCI--	PURS-sih	Perch Fish	Percidae, Percoid, Perch, Percoidious

ROOT/PREFIX/SUFFIX	PRONUNCIATION	MEANING	EXAMPLES OF USAGE
PEREGRIN--	PEREHgrin	Travel from Place to Place	Peregrinate, Peregrine falcon, Peregrination
PERENNI--	pehRIN-nih	Enduring (for at least one year)	Perennial, Perennially, *Perenni branchiata*
PERFIDE	purFIDDY or PURfid-ee	Treachery	Perfidious, Perfidy, Perfidiousness
PERFOR--	PERFor	Bore Through	Perforation, Perforate, Perfoliate
PERFUSE	perFYOOZ	Pour Through	Perfusion, Perfusate, Perfuse
PERI--	PERih	Around	Perimeter, Pericardium, Peritoneum
PERIANTH--	PERRYanth	Flower-covering	Perianthium, Periantheous, Perianthial
PERIOD	PERI-ud or PEERih-ud	a Portion of Time	Periodic, Periodate, Periodicity
PERIOST--	PERih-ahst or PERih-oh	Around a Bone	Periosteum, Periosteal, Periostracum
PERISH	PEHRish	Go Through (die)	Perish, Imperishable, Perishability
PERISSO--	purRICE-oh	Unequal (in number)	Perissodactyl, Perissodactylism
PERPEND--	perPEND	Through-weigh	Perpendicular, Pendulum, Perpend
PERPET--	perPET	Through-seek	Perpetual, Perpetuate, Perpetuation
PERSECU--	PERSehKEW	Through-follow	Persecution, Persecute, Persecutor
PERSEVE--	PERSehveh	Through-strict	Persevere, Perseverance, Perseverate
PERSIST	perSIST	Through-stand	Persistent, Persist, Persistently
PERVER--	PURvur	Turning (away from the right way)	Pervert, Perversion, Perverse, Perverted

ROOT/PREFIX/SUFFIX	PRONUNCIATION	MEANING	EXAMPLES OF USAGE
PESSIMI--	PESsim-ih	Worst	Pessimism, Pessimistic Pessimist
PEST	Pest	Plague	Pester, Pestiferous, Pestilential
PETAL--	PETl	a Leaf	Petalism, Petalody, Petaliferous
--PETE--	PETeh	Seek	Centripetal, Petition, Petasus
PETER--	PETr	Rock	Petrify, Petroleum, Petrology
PETIOL--	PETTYōl	Fruit-stalk	Petiolar, Petiolate, Petiolaceous
PETUL--	PETyool	Attack	Petulant, Petulancy, Petulantly, Impetuous
--PHACO--	FAHK-oh	Lens, Lentil	Phacoid, Angiophacus, Phacocele
PHAENO	FEEN-oh, FAY-no, FINE-oh	Show	Phaenogram, Phaenogamia, Phaenogamous
--PHAGO--	FAY-go	Eating	Phagocyte, Phagedaena, Phytophagous
PHAL--	FAL (as in "pal")	Finger	Phalanges, Phalanx, Phalanger
PHALIS	FALiss	Penis (image of)	Phallus, Phallic, Phallicism
PHAN--	Fan	Show, Appear	Phantom, Cellophane, Diaphanous
PHANERO--	FANNER-oh	Visible, Apparent	Phanerocrystalline, Phanerogamia, Phanerogamous
PHARMAC--	FARMus	Drugs, Medicine	Pharmaceutical, Pharmacy, Pharmacology
PHARYNG--	FAIRinj	Throat	Pharyngeal, Pharyngoscope, Pharyngotomy
--PHASI	FAYzee, FAHsee	Speech	Aphasia, Aphasy, Aphasic, Dysphasia

ROOT/PREFIX/SUFFIX	PRONUNCIATION	MEANING	EXAMPLES OF USAGE
PHENO--	FEEno	to Shine	Phenol, Phenocryst, Phenolphthalein
PHENO--	FEEN-oh	Show	Phenomenon, Phenomenal, Phenotype
--PHILO--	FILLoh or FILEoh	Love	Philosophy, Philanthropy, Pedophile
PHLEBO--	FLEE-bo	Vein	Phlebitis, Phlebosclerosis, Phlebotomy
PHLEG--	Fleg	Flame, Burn	Phlegm, Phlegmasia, Phlegmatic, Phlox
PHLOB--	FLOWb	Tree-bark	Phlobophene, Phloem, Phlogopite
--PHOBIA	FŌbee-ah	Madness, Fear	Claustrophobia, Hydrophobia, Acrophobia
--PHON--	Phone or Fahn	Sound	Phonetic, Telephone, Phonograph
--PHORE	Fore	Bear, Bearer	Gametophore, Semaphore, Cytophore
--PHORIA	FOREE-ah	Bearing	Euphoria, Esophoria, Phoronomy
PHOS--	FAHss	Light	Phosphine, Phosphorescent, Phosphorus
PHOTO--	FOE-toe	Light	Photograph, Photon, Photosynthesis
--PHRAGM--	Fram (as in "ham")	Fence	Diaphragm, Phragmocone, Phragmospore
PHRAS--	Fraze	Speech	Paraphrase, Phraseology, Phrase
PHREN--	Fren (as in "when")	Mind, Nerves	Phrenitis, Phrenesiac, Phrenalgia, Frenzy
PHTHEIR	fTHEER	Lice, Lousy	Phtiriasis, *Phthirus pubis*, Phthirophagous
PHTHIS--	fTHISs	Waste Away, Shrink	Phthisis, Phtisic

ROOT/PREFIX/SUFFIX	PRONUNCIATION	MEANING	EXAMPLES OF USAGE
PHYCO--	FIKE-oh	Seaweed, Algae	Phycocyanogen, Phycomycetes, Phycology
--PHYL--	File	Tribe, Class	Phylum, Phylogeny, Organophylic
--PHYLL--	Fill	Leaf	Chlorophyll, Megaphyllous, Phyllomorph
PHYSIC--	FIZZik	Nature	Physical, Physician, Physic
PHYSIO--	FIZZY-oh	Produce	Physiocracy, Physiognomy, Physiography
--PHYSIS--	FISSiss	Change, Growth	Physique, Epiphysis, Hypophysis
PHYTO--	FIGHT-oh	Plant	Phytotoxic, Phytochemistry, Phytogenesis
PIANO	pYAHN-oh	Soft	Pianissimo, Pianist, Piano, Pianette
PIEZO--	pYEZZ-oh	Press, Pressure	Piezoelectric, Piezochemistry, Piezomometer
PILA	PILLah	Little Ball	Pilule, sleeping Pill, Piles (hemorrhoids)
PILE	Pile	Cap	Pileated, Pileate, Pileum
PILUS	PILLus or PILE-us	Hair	Piliferous, Pilous, Pilar, Pile
PINNA--	PINnah	Feather, Wing	Pinion, Pinnate, Pinnately, Pinnigrade
PISCI--	PISsee	Fish	Piscatory, Pisces, Piscivorous
PLACENT	plahSENT	Cake, Plate	Placenta, Placentation, Plaque
PLACEO--	PLAHsee-oh	Pacify, Quiet	Placebo, Placate, Placid, Placative
PLAGIA--	PLAYjee-ah	Kidnapping	Plagiarize, Plagiarism, Plagiarist

ROOT/PREFIX/SUFFIX	PRONUNCIATION	MEANING	EXAMPLES OF USAGE
PLAGIO--	PLAHjee-oh	Sidewise	Plagal, Plagioclase, Plagiotropism
PLAGUE	PLAYg	Blow, Distress	Plague, Plaguey, Bubonic plague
--PLAIN--	PLAYn	to Beat the Breast	Complaint, Complain, Plaintiff
PLANE--	PLANeh	Wanderer, Planet	Planet, Planetary, Planetarium, Planoblast
PLANO--	PLAY-no or PLANno	Flat, Moving	Plano-convex, Planogamete, Planometer
PLANTA	PLANtah	Plant	Plantation, Plantain, Plantlet
PLANUS	PLANnus	Flat	Plane, Planarian, Plan, Plantigrade, Planula
--PLASM--	PLAZm	Molded Image, Molded Form	Plastic, Cytoplasm, Plasmolysis
--PLAST--	Plast	Formed	Leucoplast, Cytoplast, Plastogamete
--PLASTY	PLASTih	Growth, Formed	Thoracoplasty, Autoplast, Neoplasty
PLAT	PLAYt	Flat	Platitude, Platform, Plate glass
PLATY	PLATtih	Flat, Broad	Plateau, Platyhelminthes, Platypus
PLAUD	Plawd	Applaud	Plausible, Plaudit, Laudable, Implausible
PLEB--	Pleb or Pleeb	the Common People	Plebiscite, Plebeian, Plebeianism
--PLECT, PLEX	PLEKt, PLECKs	to Braid, Weave	Complex, Complexus, Complexion
PLENA--	PLENnah, PLEENnah	Full	Plenary, Plenitude, Plenipotentiary
--PLEO--	PLEE-oh	to Fill	Complete, Replete, Deplete, Complement
PLEO--	PLEE-oh	More	Pleochroic, Pleochromatic, Pleomorphism

ROOT/PREFIX/SUFFIX	PRONUNCIATION	MEANING	EXAMPLES OF USAGE
PLETH--	PLEHth	Too Full	Plethora, Plethoric, Plethysmograph
PLEUR--	Plewr	the Side	Pleura, Pleurisy, Pleuritic
PLEX--	Plecks	to Strike	Plectrum, Plectron, Plexor, Pleximeter
PLIA--	PLYah	to Fold	Pliant, Compliance, Plica, Plication
PLUMB	PLUMb	Lead (metal)	Plumber, Plumb, Plumbeous
PLUME	Ploom	a Feather	Plumose, Plumosity, Plume, Nom de plume
PLUTO	PLOOT-oh	Wealth	Plutocracy, Plutolatry, Plutologist
--PLUVI--	PLOOVih	Rain	Compluviam, Pluvial, Pluviometer
PNEUMO--	NEWmō	Air, Breath	Pneumococcus, Pneumatophore, Pneumatic
--POD--	PAHd or Pōd	Foot	Tripod, Hexapod, Chiropody, Podium
PODO--	PŌ-dough	Foot	Podobranchia, Podium, Podophyllin
POEM--	PŌem	Make	Poesy, Poem, Poetry, Poetic
POIKILO--	POYK-L-oh	Change, Mottled	Poikilothermic, Poikiloblast, Poikilocyte
--POLA--, --POLI--	Pole-ah, POLE-ih	Polish, Refurbish	Interpolate, Extrapolate, Polish
POLAR--	POLEur	Be in Motion	Polar, Polaris, Polarization, Polarize
--POLI--	PŌ-li	City	Metropolis, Police, Political
POLLEN	PAHLin	Fine Flour	Pollinate, Pollen, Pollinization

ROOT/PREFIX/SUFFIX	PRONUNCIATION	MEANING	EXAMPLES OF USAGE
POLY--	POLLY	Many	Polygamy, Polygon, Polymorphism
POLYP--	PAHLip	Many-footed	Polypary, Polypod, Polypus, Polyp
POMA--	PŌmah	Apple	Pomace, Pomaceous, Pomade, Pome
POMUM	PŌmum	Fruit	Pomona, Pomiferous, Pomegranate
--PONO	PO-no	Place (in relation)	Expose, Proponent, Opponent, Exponent
POPU--	POPyoo	People	Populate, Population, Popular
PORE	Pōre	a Tiny Passage	Poriferous, Porosity, Porifera, Porism
PORNO--	PORN-oh	Prostitute, Sell	Pornographic, Pornerastic, Pornography
PORPHYR	POR-fir	Purple	Porphyratin, Porphyria, Porphyry
PORT--	Port	to Carry	Seaport, Portable, Portage, Portfolio
PORTA--	Port-tah	Gate	Portal, Portcullis, Porch
--POSI--	POsih	Place	Position, Deposit, Imposition
POSSE	PAHSSee	Be Able	Posse, Possible, Possess
POST--	Post	After	Post mortem, Postglacial, Posterior
POSTUL--	PAHSTyool	Demand, Require	Postulant, Postulate, Postulation
POT, POTA--	Paht, PŌtah	River, Drink	Potable, Potion, Hippopotamus, Potatory
POTEN--	PŌtn, PŌ-ten	Be Able	Potent, Potentiate, Potentiality
PRACTI--	PRACKTih	Do	Practice, Practical, Practicable, Praxis

ROOT/PREFIX/SUFFIX	PRONUNCIATION	MEANING	EXAMPLES OF USAGE
PRAGMAT--	prahg-MAHT or pRAHG-mat	Versed in Practical Affairs	Pragmatic, Pragmatist
PRE--	Pree	Before (in time, rank, or position)	Predigest, Preamble, Predict, Preclude
PRECIP--	PRESsip	Headlong Fall	Precipitate, Precipice, Precipitant
PREDA--	pRED-ah or pREEDah	Plunder	Predaceous, Predatory, Predator
PREDICA--	PREDIKah	Before-proclaim	Predicament, Predicate, Predictable
PREHEND	preHEND	to Seize	Comprehend, Apprehend, Reprehensible
PRESBY--	PREZbe	Old Man	Presbyopia, Presbyterate, Presbycusis
PRETER--	PRETTer	Beyond-go	Preterite, Preterition, Preternatural
PRIMA--	PREEma or PRĪMal	First	Primal, Primitive, Primacy, Primary
PRISM	PRIZm	Saw	Prismatic, Prism, Prismoidal
PRO--	Prō	For, Yes	Proton, Proportion, Pro rata, Pronoun
PRO--	Prō	Before, First	Prognosis, Proboscis, Prologue, Project
PROBO--	PRŌB-oh	Test, Prove	Probe, Probable, Proboscope, Probate
PROCED--	prōSEED	Before-go	Procedure, Procedendo, Proceed
PROCTO--	PRŌK-toe or PRAHKT-oh	Anus	Proctoscope, Proctodaeum, Proctology
PRODUC--	PROduck	Before-lead	Productive, Produce, Producer

ROOT/PREFIX/SUFFIX	PRONUNCIATION	MEANING	EXAMPLES OF USAGE
PROFESS	proFESS	Forth-confess	Professor, Profession, Professional
PROFUS--	proFYOOZ	Forth-pour	Profusion, Profuse, Profusely
PRONA--	PROnah	Bend forward	Pronaos, Pronate, Pronation, Pronator
PRONOUN	PROnown	for a Word, Announce	Pronoun, Pronounce, Pronominal
PROPAG--	PROpahg	Multiply, Generate	Propagate, Propagation, Propaganda
PROPEL	proPELL	Forward-drive	Propellant, Propeller, Propel, Propulsion
PROSTATE	PRAHST-ate	One in the Front Ranks	Prostate gland, Prosthesis, Prosthetic
PROTEC	proTEK	Before-cover	Protective, Protection, Protege
PROTEST	PROtest or proTEST	Before-witness	Protestation, Protestant, Protestingly
PROTO	PRO-TOE	First	Protoplasm, Prototype, Protozoa
PROTRACT	proTRAKT	Forward-draw	Protracted, Protractor, Protraction
PROVID--	PROvid	Forward-see	Providence, Provide, Provisions, Proviso
PROVOC--	PRO-VOKE, PRO-VAHK	Forth-call	Provocative, Provoke, Provocation
PROX--	PRŌCKS, PRAHKs	Near, Next	Proximal, Proxy, Proximo, Approximate
PRUD--	Prood	Providential, Good	Prudence, Prudent, Prudential, Prude
PRUNE	Proon	Plum	Pruniferous, Prunella, Prunello, Prune
PRURI--	PROOrih	Itching	Prurient, Prurience, Prurigo, Pruritis
PRYTAN--	PRITTan	Foremost-perhaps	Prytaneum, Prytanis, Prytany

ROOT/PREFIX/SUFFIX	PRONUNCIATION	MEANING	EXAMPLES OF USAGE
PSEUDO--	SOO-dough	False	Pseudonym, Pseudopodia, Pseudomorph
PSITTA--	SITtah	Parrot, Bird	Psittacosis, Psittaci, Psittacean, Psittacid
PSORA	SORah	Itch, Scurvy	Psorlea, Psoriasis, Psorosis, Psoralen
PSYCHE--	SIGH-key	Mind, Spirit	Psychology, Psychic, Psychotic
--PTER--	Tehr or Turr	Wing, to Fly	Helicopter, Apterous, Pterodactyl
PUBER--	PYOOber	of Ripe Age	Puberty, Puberous, Pubescent, Pubic
PUBLI--	PUBlih	People, Populace	Public, Publication, Popular, Publicize
PUDEN--	PYOODn	Bashful	Pudency, Pudendum, Pudendal
PUERP--	pyooURP	Child	Puerperal, Puerpera, Puerperium
PUGI--	PYOOjih	Fight	Pugnacious, Pugilism, Pugilist, Repugnant
PULLA	PULLah	Young	Pullet, Pullulate, Pullulation, Pullulative
PULMO--	PULmo	Lung	Pulmonary, Pulmometer, Pulmotor, CPR
PULS--	PUHLss	Beat	Pulsation, Pulsate, Pulsimeter, Pulse
PULV--	PUHLv	Powder, Grind, Dust	Pulverize, Pulverization, Pulverulent, Powder
PULVI--	puhlVIN	Cushion, Swelling	Pulvillus, Pulvinar, Pulvinate, Pulvinus
PUNCT--	Punkt	Point, to Punch	Punctual, Punctate, Puncheon, Punctuate
PUNI--	PYOONih	Punish	Punishable, Punitive, Punishment
PUPA--	PYOOPah	Girl, Doll	Pupae, Pupation, Pupate, Puppet

ROOT/PREFIX/SUFFIX	PRONUNCIATION	MEANING	EXAMPLES OF USAGE
PURG--	Purj or Purg	to Clean	Purge, Purgatory, Purgative, Purging
PURI--	PYOORih	to Make Pure	Puritan, Purify, Purification, Purity
PURPU--	purPYOO	Purple, Shell	Purpura, Purpurate, Purpuric acid
PURUL--	PYOORrule	Pus	Purulent, Puruloid, Purulence, Pustule
PUTRI, PUTRE--	PYOOTrih, PYOOTreh	Rotten	Putrefy, Putrid, Putrefaction, Putrescence
PYCNO--, PYKNO--	PICKno	Thick	Pycnidiospore, Pycnidium, Pycnis, Pycnium
PYELO--	PȲel-oh	Pelvis, Kidney	Pyelitis, Pyelocystitis, Pyelostomy
PYGO, PYGI--	PIGgo, PIDJee	Rump, Buttocks	Pygidium, Pygal, Pygodidymus, Pygomelus
PYLO--	PIE-lō	Gate, Valve	Pylon, Pyloric, Pylorus, Pyloralgia
PYO--	PIE-oh	Pus, Pus-former	Pyogenesis, Pyorrhea, Pyoderma
PYRE--	PIEr	Fever	Pyretic, Pyrexia, Pyretogenesis
PYRENO, PYREN--	pieREENoh, PIEren	Fruit-stone	Pyrene, Pyrenoid, Pyrenodean
PYRO--	PIE-ro	Fire, Heat	Pyrogen, Pyrolysis, Pyromaniac
PYTHO--	PIEtho	Filth, Rot	Pythogenesis, Pythogenous, Pythogenetic
PYX--	Picks	Box	Pyx chest, Pyxis, Pyxidium, Pyxides
QUAD--	KWAHD	Four	Quadrangle, Quadrilateral, Quadruped
QUALI--	KWAHLih	Quality	Qualify, Qualitative, Quality

ROOT/PREFIX/SUFFIX	PRONUNCIATION	MEANING	EXAMPLES OF USAGE
QUANT--	KWAHNT	How Much	Quantify, Quantitative, Quantity, Quantum
QUAR, QUER--	kWAR, KWEHr	Complains	Quarrel, Quarrelsome, Querulous
QUART	Kwart	Four, Fourth	Quartan, Quartation, Quarter, Quartet
QUASI--	KWAHsih	As If, Just As	Quasi-contract, Quasi-judicial, Quasar
QUATERN	KWATturn	Four	Quaternary, Quaternion, Quatrain
--QUER, QUES--	Kwhere, Kwess	to Seek	Inquest, Question, Quest, Conquest
QUERCIN	KWERsin	Oak	Quercine, Quercetin, Quercitron, Quercus
QUIET	KWĪet	Quiet, Rest	Quiescent, Quiet, Quietus, Quietude
QUIN--	Kwin	Five	Quintuplets, Quinate, Quincuncial
QUINQUE--	KWINkwah	Five	Quinquennial, Quinquevalent, Quinquefoliate
QUOTA	KWŌtah	How Many	Quotennial, Quotidian, Quotient, Quota
RACEM--	raySEEM or RAYsem	Cluster (of grapes)	Raceme, Racemiferous, Racemic acid
RACHIS	RAYkiss	Spine	Rachilla, Rachis, Rachitic, Rachitis
RADIA--	RAYdee-ah	To go out from (as a light ray)	Radiate, Radial, Radio, Radioactive
RADIX	RADdix	Root	Eradicate, Radical sign, Radish
RADULA	RAJYOOlah	Scrape	Radulated, Radular, Radulate
RAGE	RAYj	Rave, Rage	Rabid, Rabies, Raging, Enrage

ROOT/PREFIX/SUFFIX	PRONUNCIATION	MEANING	EXAMPLES OF USAGE
RAMI--	RAMmih	to Branch	Ramiform, Ramification, Ramify, Ramose
RANA	RAHNah	Frog	Ranarian, Ranarium, Ranidae, Rana
RANC--	RANk or RANss	Stinking	Rancid, Rancor, Rancorous
RANU--	RANnoo	Little Frog	Ranunculus, Ranunculaceous, Ranarium
RAPH--	RAYfee	Sew, Stitch	Raffia, Raphe, Raphide, Raphis
RAPI--	RAPih, RAYpih	Plunder	Rapid, Rapine, Rapaceous, Ravish
RAPT--	Wrapped	to Seize	Raptor, Rapt, Rapture, Rape
RARUS	RARE-us	Rare	Rarity, Rarely, Rarefaction
--RAS--	Rass	Scrape	Razor, Rasp, Abrasion, Abrade
RASC--	Rask	Rash	Rascal, Rascality, Rash
RATI--	RATtih	Fixed	Ratify, Rated, Ratio, Ration
RATIO--	RASHya	Reason	Rationalize, Rationally, Rational
RATIT--	RATtit	Raft-like	Ratitae, Ratite, Ratis
RECALCIT--	re-KALSsit	Kick Back with the Heels	Recalcitrant, Recalcitrate, Recalcitrance
RECAPIT--	re-kahPIT	Again-Head (start again at the heading)	Recapitulation, Recapitulate, Recapitulatory
RECED, RECES--	reSED, reSESS	Conceal, Recede	Recess, Recession, Recessive
RECEIV	reSEEV	Take-back	Receive, Reception, Receptor
RECIDI--	reSIDDih	Fall Back	Recidivist, Recidivism, Recidivous

ROOT/PREFIX/SUFFIX	PRONUNCIATION	MEANING	EXAMPLES OF USAGE
RECIPR--	re-SIPPr	Mutual, Alternating	Reciprocate, Reciprocal, Reciprocity
RECLINE	reKLINE	Lean Back	Recline, Reclination, Reclinate
RECOGNI--	reKAHGnih	Know-again	Recognition, Recognizable, Recognize
RECONCILE	REKahn-sīle	Council Again	Reconcilable, Reconciliation, Reconciler
RECT--	Wrecked	Right, Straight	Rectangle, Erect, Rectilinear, Rectify
REFLECT	reFLECKt	Bend Back	Reflection, Reflective, Reflex
REFLU--	REFFloo	Again-flow	Refluent, Refluence, Reflux
REFRACT	reFRAKT	Back-break	Refraction, Refractive index, Refractor
REGULA--	REGyoo-lah	Rule	Regulate, Regulus, Regulator
REGURGIT--	re-GURJit	Again-abyss (from the gorge)	Regurgitant, Regurgitate, Regurgitation
REIGN, REGN--	RAIN, REGn	Rule	Reign, Regal, Regnant, Regent
REJECT--	reJEKT or REEjekt	Back-throw	Rejection, Rejectamenta, Rejecter
RELAT--	REHlut	Borne-back	Relativity, Relation, Relative, Relate
RELAX	reLACKs	Again-loose	Relax, Relaxation, Relaxable, Release
RELIC	RELL-ik	Leave-back	Relict, Relic, Derelict, Dereliction
RELIEV	reLEEV	Again-light (hearted)	Relief, Relieve, Alleviate
RELUCT	reLUCKt	Strive-against	Reluctant, Reluctance, Reluctivity
REMEDY	REHMed-ee	Heal-again	Remedial, Remedies, Remediate

ROOT/PREFIX/SUFFIX	PRONUNCIATION	MEANING	EXAMPLES OF USAGE
REMEX	RE-meks	a Leading Oar (a wing quill)	Remigial, Remiges, Remex
REMONSTRATE	reMAHN-straight	Again-show	Demonstrate, Remonstrance, Remonstrate
REMORSE	rehMORSE	Bite One's Back	Remorse, Remorseful, Remorseless
REMOTE	rehMOAT	Removed	Remotion, Remote control, Remotely
REN--	Wren	Kidney	Renifleur, Reniform, Renin, Renal
RENAISSANCE	RENnah-SAHNss	Born Again	Nascent, Renascence, Neonate, Renascent
RENDER	RENDr	Give-back	Rendition, Render, Rendering, Surrender
RENOVAT	RENo-VATE	New-again	Renovate, Renovation, Renovative
REPAIR	rehPARE	Make Ready Again	Irreparable, Repairs, Reparations
REPEAT	rePEET	Again-seek	Repeated, Repetend, Repeating decimal
REPLY, REPLI--	reePLY, REPP-Lih	Fold, Again	Replicate, Replication, Reply, Replica
REPRESENT	rehPREEzent	Again-Present (a copy for the mind)	Representation, Representative, Represent
REPTA--, REPTI--	REPTah, REPTih	to Creep	Reptile, Reptant, Reptilian
REPUDI--	rePYOODih	Divorce	Repudiate, Repudiation, Repudiative
REPUG--	rePUGG	Fist-back	Repugnant, Repugnance, Pugnacious
REPUTE	rehPYOOT	Think-again	Reputation, Reputable, Reputedly
REQUE, REQUI--	Wreck, REKwih	Seek-again	Requisite, Require, Request, Requisition

ROOT/PREFIX/SUFFIX	PRONUNCIATION	MEANING	EXAMPLES OF USAGE
RERE	Rear	Rear	Arrears, Reredos, Rere-mouse, Rere
RESENT	reZENT	Feel-against	Resentment, Resent, Resentful
RESID--	REZid or RESsid	Reside, Remain	Residue, Residence, Residual, Residuum
RESIL--	reSILL	Leap Back	Resilient, Resiliometer, Resilience
RESIN	RESsin	Resin	Rosin, Resinous, Resinoid, Resinate
RESIST	reZIST	Set Back	Resistance, Resistant, Resistor
RESOLVE	reZAHLv	Loosen Again	Resolvable, Resolvent, Resolving power
RESONA	REZ-o-na	Sound Again, Echo	Resonance, Resonator, Resounding
RESPIRE	re-SPIRE	Breath Back	Respiration, Inspire, Respirator
RESPOND	reSPAHND	Again Promise	Respondent, Response, Correspondent
RETEN--	reTEN	Hold Back	Retain, Retention, Retentive, Retinue
RETI--	RETtih	a Net	Retina, Retiary, Reticle, Reticulate
RETI--, RETIN--	RETti, RETtin	Resin	Retinite, Retinol, Retinitic
RETOR--	reTOR	Back-twist	Retort, Retorsion, Retortion
RETRAC--	reTRACK	Back-pull	Retraction, Retractile, Retractive, Retreat
RETRIBU--	rehTRIB-yoo	Again-allot	Retribution, Retributive, Retributor
RETRO--	RETtro	Backward	Retrograde, Retrogress, Retrovert
RETROSPECT	RETro-spekt	Look Backward	Retrospect, Retrospective, Retrospection

ROOT/PREFIX/SUFFIX	PRONUNCIATION	MEANING	EXAMPLES OF USAGE
RETURN	reTURN	Back + Round-off	Returnable, Return, Returns, Returned
REVEA, REVE--	reVEE, REVeh	Throw Back the Veil	Reveal, Revealable, Revelation
REVERBER--	reVERBr	Again-whip	Reverberate, Reverberant, Reverberations
REVERE	REVer-eh	Again-fear	Reverence, Reverend, Reverent, Venerate
REVERS--	reVURSS	Back-turn	Reversible, Reversion, Revert
REVI--	REHVih or REEVih	Again-see	Review, Revise, Reviewal, Revision
REVOC, REVOK--	reVŌKE	Again Call	Revoke, Revocation, Revocable
REVOL--	rehVAHL or REVahl	Back-roll	Revolver, Revolve, Revolution, Revolt
REX	Wrecks	King	Regal, Regnant, *Tyrannosaurus rex*
RHABD--	RABd	Rod	Rhabditis, Rhabdomancy, Rhabdocoele
--RHAGE, RHAGIA	Rage, RAGah	Break	Hemorrhage, Rhagades, Osteorrhagia
RHAP, RAPH--	RAP, RAFF	Sew, Seam	Rhapsody, Raphe, Raphide
--RHEA	REEah	Flowing	Diarrhea, Gonorrhea, Menorrhea
RHEO--	REEoh	Flow	Rheostat, Rheoscope, Rheotaxis
RHETOR	REHTtor	Orator, Speaker	Rhetoric, Rhetorical, Rhetorician
RHEU--	ROO	Flow	Rheum, Rheumatism, Rheumatoid
--RHIG--	Rig	Frost, Cold	Rhigolene, Frigid, Frigorific

ROOT/PREFIX/SUFFIX	PRONUNCIATION	MEANING	EXAMPLES OF USAGE
RHINO--	RHINE-oh	Nose	Rhinoceros, Rhinovirus, Rhinoplasty
RHIZO--	RISE-oh	Root	Rhizome, Rhizopod, Rhizocarpus
RHOD--	Road	Rose	Rhodamine, Rhodium, Rhododendron
RHOMB--	RAHMb	Revolve	Rhombic, Rhombiform, Rhomboid
RHYPAR--	RIPpar	Filthy	Rhyparography, Rhyparographer, Rhyparographic
RIBO--	RYE-bo	a Sugar	Riboflavin, Ribose, Ribonucleic acid
RICI--	RISSih	Castor Oil Plant	Ricin, Ricinoleic acid, Ricinolein
RID--	Rid	Laugh, Smile, Jest	Riant, Rident, Ridicule, Ridiculous
RIGID	Ridge-id	Stiff	Rigidity, Rigor, Rigorous, Rigor mortis
RIM--	Rhyme	Chink, Cleft	Rime, Rimose, Rimous, Rimosity
RINGO--	RIN-go	Gape	Rictus, Ringent, Ringose
RIPA--	RIPah	River Bank	Riparian, Riparious, Ripuarian, River
RITE, RITU--	Rite, RITyoo	Rite	Ritual, Rite, Ritualistic
ROBUS, ROBUR	rōBUS, RŌbus	Strength	Robust, Roborant, Roberite
ROGO	ROgo	Ask, Question	Rogation, Interrogate, Interrogative
ROSTRUM	RAHstrum	Beak	Rostellum, Rostrad, Rostral, Rostrate
ROTA, ROTO	RŌtah, RŌtoe	Wheel, Circle	Rotor, Rotary, Rotifera, Rotation
RUBE--	ROOBeh	Red, Rosy	Ruby, Rubeola, Rubescent, Rubric

ROOT/PREFIX/SUFFIX	PRONUNCIATION	MEANING	EXAMPLES OF USAGE
RUMI, RUME--	ROOMY, ROOMeh	Throat	Ruminant, Ruminate, Ruminantia, Rumen
--RUPT, ROMP--	RUHpt, ROAMp	Break	Rout, Rupture, Interrupt, Abrupt
RURAL	ROORal	Country, Farming	Rustic, Rural, Ruralism, RFD
SABU--	SAByoo	Sand, Gritty	Sabulous, Saburral, Sabulosity
SAC--	Sack	Pouch, Sack	Sack, Saccate, Sacculus
SACCHAR	SAHKcar	Sugar	Saccharide, Saccharine, Saccharoidal
SACR--	SAKEr, SACKr	Sacred, Holy	Sacred, Sacrifice, Sacrament, Sacrum
SAGAC--	sahGASS	Perceive Quickly	Sage, Sagacious, Sagacity
SAGITA--	SAJ-itah	Arrow	Sagittal, Sagittarius, Sagittary, Sagittate
SAL--	Sal	Salt	Sal ammoniac, Salary, Saline
SALAC--	SALlus	Venal, to Sell One's Honor	Salable, Salacious, Salaciousness
SALIC--	SALiss	Willow	Salicaceae, Salicin, Salicylic acid
SALIV--	SALIHv	Spittle	Salivary, Salivate, Saliva, Salivation
SALPA--	SALpah	a Seafish	Salpidea, Salsify, Salpiform, Salpid
SALTA--	SAHLTah	Leap, Dance	Saltation, Saltant, Salticidae
SALU--	SALoo	Health	Salute, Salubrious, Salutiferous
SALV--	Salve	Whole, Safe	Salvation, Salve, Salvage, Salvo
SANCT--	SANkt	Holy	Sanctity, Sanctimonious, Sanctuary

ROOT/PREFIX/SUFFIX	PRONUNCIATION	MEANING	EXAMPLES OF USAGE
SANGUI--	SAHNgwee	Blood	Sanguine, Sanguicolous, Sanguiniferous
SANIT--	SANnit	Heal	Sanitary, Sanitive, Sanitation
SANIT	SANnit	Health	Sanitarium, Sanitize, Sanitarian
SAPID--	SAPid, SAYpid	Taste, Flavor	Insipid, Sapidity, Sapidness, Savor
SAPIEN	SAYpee-en	Wise	Sapient, *Homo sapiens*, Savant
SAPO--	SAPoh	Soap	Saponify, Saponic acid, Saponin
SAPRO--	SAPRo	Rotten	Saprophagous, Saprophyte, Saprogenic
SARCAS--	SARKus	Tear-flesh	Sarcasm, Sarcastic, Satire
SARCO--	SARK-oh	Flesh	Sarcophagidae, Sarcophagus, Sarcoma
SARMENT--	sarMENT	Prune, Runners	Sarmentum, Sarmentose, Sarmentaceous
SARTOR	SARTor	Tailor, Patcher	Sartorial, Sartorius, Sartor
SATI--	SAT-ih	Fill up, Enough	Saturate, Satisfy, Satiety
SATYR	SATtr	Goat-like, Excessive Male Sexuality	Satyriasis, Satyric
--SAUR--	SAWur	Lizard	Saurian, Dinosaur, Ichthyosaur
SAXI--	SACKSih	Stone, Rock	Saxifrage, Saxicholine, Saxitoxin
SCABI--	SCABbih	Rough, Itch	Scabrous, Scabies, Scabious, Scab
SCALA--	SKAHLah	Staircase, Ladder	Escalade, Scalar, Scalariform
SCALEN--	SKAYlen	Uneven	Scalenous, Scalene triangle, Scalenohedron

ROOT/PREFIX/SUFFIX	PRONUNCIATION	MEANING	EXAMPLES OF USAGE
SCAND--	Scanned	to Climb	Scandent, Scandic, Scansorial, Scan
SCAPH	SKAFf	Hollow, Bowl	Scapha, Scaphoid, Scaphopod
SCAPULA--	SKAP-YOO lah	Shoulder	Scapular, Scapulalgia, Scapula
SCARAB	Scare-ub	Beetle	Scarabaeidae, Scarabidous, Scaraboid
SCATO--	Scat	Feces, Dung	Scatology, Scatoma, Scatoscopy
SCENE	SEEn	Stage	Scenic, Scenery, Scenography
SCHEMA--	SKEEMah	Form	Scheme, Schematic, Schematically
SCHISM	SHIZm (often SKIHZm)	Split, Cleave	Schizophrenia, Schizomycetes, Schist
SCHOLAR	SKAHLar	School	Scholastic, Scholium, Interscholastic
SCIEN--	SIGH-en	to Know	Scientist, Science, Scientific
SCINTIL--	SINTtil	Light	Scintilla, Scintillate, Scintillation
SCIO, SCIA--	SKEEoh, SIGHah	Shadow, Shade	Skiagram, Sciophilous, Sciophyte
SCIRRH--	SKEER	Tumor	Scirrhus, Scirrhoid, Scirrhosity
SCISS--	SIS	to Cut	Scission, Scissile, Scissors
SCLERO--	SKLEHR-oh	Hard	Scleroderma, Sclerosis, Sclerata
SCOLIO--	SKŌLee-oh	Curved, Twisted	Scoliosis, Scoliotic, Scoliokyphosis
--SCOP	SKŌP	See	Telescope, Periscope, Microscope

ROOT/PREFIX/SUFFIX	PRONUNCIATION	MEANING	EXAMPLES OF USAGE
--SCOPY	SKAHPee	Observation	Enteroscopy, Laryngoscopy, Proctoscopy
SCORBUT--	SKORbut	Scurvy	Scorbutic, Scorbutigenic, Scorbutus
SCORPIO--	SKORpee-oh	Scorpion	Scorpioid, Scorpion fly, Scorpion
SCOUR	SKOU-wer	to Run Out	Scurge, Scour, Scoundrel, Scourings
--SCRIB, SCRIP--	Skrib, Skrip	Write	Script, Describe, Inscription, Prescribe
SCROB--	SKRŌb	Marked with Pits	Scrobiculate, Scrobiculus, Scrobiculated
SCROFU--	SKRAWFyoo	Breeding Sow	Scrofula, Scrofuloderma, Scrofulous
SCROTUM	SKRŌt-um	Pouch Holding the Testes	Scrotitis, Scrotectomy, Scrotocele
SCRUPLE	SKROOpl	Minute Weight, Sharp Stone	Scrupulous, Unscrupulous, Scrupulosity
SCRUTA--	SKREWtah	Search Carefully	Scrutinize, Inscrutable, Scrutiny
SCULP--	SKULp	to Carve	Scalpel, Sculpture, Sculptor
SCUTUM	SKYOOtum	Shield	Scutellum, Scutate, Scuttle, Scutella
SCYPHO, SCYPHI--	SIGHfō, SIGHfih	Goblet, Cup	Scyphiform, Scyphozoan, Scyphozoid
SEBUM	SEEBum	Suet, Oily	Sebaceous, Sebacic, Seborrhea, Sebolith
SECLUD	sehKLOOD	Aside-shut	Seclusion, Secluded, Seclusive
--SECRET--	SEEKret or sehKREET	Aside-separate	Secret, Secretion, Secretory, Excrete
--SECT--	SEKt	Cut	Transect, Dissect, Insect, Section

ROOT/PREFIX/SUFFIX	PRONUNCIATION	MEANING	EXAMPLES OF USAGE
SEDAT--	SEDDut	Allay (pain)	Sedative, Sedation, Sedate
SEDEN--	SEDen	Sit	Sedentary, Sedentarily, Sedile, Seance
SEDIM--	SEDim	Settling	Sedimental, Sedimentary, Sediment
SEDIT--	sedDIT	Aside-going	Sedition, Seditious, Seditionary
SEDUC--	sehDUCK	Aside-leading	Seduction, Seducement, Seduce, Seductive
SEDUL--	SEDyool	Diligent	Sedulous, Sedulity, Sedulousness
SEGMENT	SEGment	to Cut	Segmentation, Segmentary, Segments
SEGREG--	SEG-greg	Aside-flock	Segregate, Segregation, Segregative
SEISM--	SIGHzm	Shake	Seismic, Seismograph, Seismology
SELACH--	si-LAKEee	Shark	Selachi, Selachian, Selachoid
SELECT	sehLEKT	Apart-pick	Selection, Selective, Select
SELEN--	sehLEEN	Moon	Selenium, Selenite, Selenographer
SEMA--	SEMma	Sign	Semantics, Semaphore, Semasiology
--SEMB, SIMB--	SEMb, SIMb	Simulate	Resemble, Semblance, Semblative, Similar
SEMI--	SEM-me, SEM-my	Half	Semiannual, Semiweekly, Semiautomatic
--SEMIN--	SEEMin	Seed	Seminar, Inseminate, Seminal Fluid
SENAT	SENnaht	Old	Senatorial, Senator, Senatus consultum
SENI--	SENnih	Old	Senile, Senility, Senior, Seniority

ROOT/PREFIX/SUFFIX	PRONUNCIATION	MEANING	EXAMPLES OF USAGE
SENSA--	SENSah	Intelligent	Sensation, Sensational, Insensate
SENSI--	SENSih	Feel	Sensibility, Sensible, Sensitive
SENSU--	SENSyoo	Pleasant-feel	Sensual, Sensuality, Sensuous
SENTI--	SENTih	Feeling	Sentient, Sentimental, Sentiently, Sentiment
SENTIN--	SENtin	Path	Sentry, Sentinel, Sentry box
SEP, SEPS--	Sep, SEPs	Lizard	Sep, Seps, Sepsina
SEPAL	SEEPul or SEPul	Covering, Apart	Sepaloid, Sepaline, Sepalous, Separate
SEPSI--	SEPsih	Make Putrid	Sepsis, Asepsis, Sepsine
SEPT--	SEHPt	Seven	September, Septuple, Septuagenarian
SEPTI--	SEPTih	Rotted	Antiseptic, Septicemia, Septic tank
SEPTUM	SEPtum	Dividing Fence	Septal, Septarium, Septate, Septum
SEQU--	SEEKwah	Result, Follow	Sequel, Sequelae, Sequence, Consequence
SEQUESTR	seeKWESTur	Lay Aside, Surrender	Sequester, Sequestration, Sequestrum
SERE, SERI--	Sear, SEHrih	Silk-like	Serge, Sericeous, Sericate, Sericulture
SEREN--	SEHRen	Clear, Bright	Serenade, Serenata, Serene
SERO	SEHR-oh	Join	Series, Serial, Seriately
SERO, SERU--	SEHR-oh, SEHRoo	Liquid, Whey	Serology, Serosity, Serum
SERPEN--	SERPpen	Creeping, Crawling	Serpent, Serpentine, Serpiginous, Serpentaria
SERR--	SERR	Lock	Serry, Seraglio, Serried

ROOT/PREFIX/SUFFIX	PRONUNCIATION	MEANING	EXAMPLES OF USAGE
SERRAT	serRAHT	Saw-edged	Serrated, Serrations, Serrulate, Sierras
SERTU, SERTA--	SERToo, SERTah	Garlands	Sertularia, Sertularian, Sertularidae
SERUS, SERO	SEARus, SER-oh	Late	Serotina, Serotinous, Serotinal
SERV--	Serve	Serve	Sergeant, Servant, Service, Servomotor
SERVU--	SERVoo	Slave	Servitude, Servile, Serf, Servility
SESQUI	SESS-kwih	One and a Half	Sesquicentennial, Sesquiplane, Sesquioxide
--SESSI--	SESSih	Sit, Perch	Session, Insessores, Insessorial, Seance
SESSIL	SESSil	Attached by a Broad Base	Sessile, Sessility, Sessiliform
SETAE	SETtee or SEEtee	Stiff Bristles	Setaceous, Setiform, Polysetaceous
SEX--	Seks	Six	Sexcuspidate, Sexennial, Sextuplets
SIALO--	SEEah-lo	Saliva	Sialagogic, Sialolith, Sialoangitis
SIBIL--	SIBill	Hiss	Sibilant, Sibilance, Sibilation
--SICC--	Sick	Dry	Dessicate, Exsiccate, Dessicant, Siccative
SIDER--	CIDERoh	Iron	Siderocyte, Siderography, Siderolite
SIDUS--	SIDE-us, sighDEAR	Star	Sideroscope, Siderial, Consider
SIGM--	SIGm	"S" shaped	Sigmoid, Sigmascope, Sigma curve
SIGNUM	SIGnum	Sign	Signify, Sign, Signal, Signate
SILEN--	SILE-en	Be Still	Silence, Silent, Silentiary, Silenced

ROOT/PREFIX/SUFFIX	PRONUNCIATION	MEANING	EXAMPLES OF USAGE
SILIC--	SILLik	Flint	Silica gel, Silicosis, Silicon
SILIQ--	silLIKE or silLEEK	Pod, Husk	Silique, Siliquose, Siliquoid
SILUR--	sigh-LOOR	Sheet	Silurian, Siluridae, Siluroid
SILVA	SILVah	Forest, Woods	Savage, Silva, Silvae, Silvanus
SIMIAN	SIMMYan	Flat-nosed	Simian, Simial, Simioid, Simious
SIMIL--	SIMmil	Similar, Alike	Similarity, Simile, Similarly, Simulate
SINAP	sinNAHP	Mustard, White	Sinalbin, Sinalbine, Sinapin, Sinapism
SINE	Sign	Curve	Sinuous, Sinuate, Sinusitis
SINISTR--	SINist-er	to the Left	Sinister, Sinistral, Sinistrorse
--SINU--	SINyoo	to Wind (in), to Bend	Insinuate, Sinuous, Sinuosity
SIPHON	SIGHfahn	Siphon	Siphonophore, Siphuncle, Siphonaptera
SIROS	SIGHrōss	a Pit for Corn	Silage, Silo, Ensilage
SITO--	SITE-oh	Grain, Food	Sitomite, Sitophobia, Sitology
SITU--	SITtoo	Place, Location	Situation, In situ, Situs, Site
SKAT--	Skat	Dung	Scatological, Scatology, Scat
SKEPTIKOS	SKEPtih-kōss	Reflective	Skeptic, Skeptical, Skepticism
SKIA--	SKYah-graf	Shadow	Skiagraph, Skiagram, Skiascope
SOL--	Soul	Sun	Solarium, Solarize, Solar eclipse
SOLEN--	SOULn	Channel	Solenoid, Solenaceous, Soleniform

ROOT/PREFIX/SUFFIX	PRONUNCIATION	MEANING	EXAMPLES OF USAGE
SOLU--	SOUL-oo or SAHLyoo	Dissolve	Solution, Solute, Insoluble
--SOLV--	SAHLv	Loose, Loosen	Solvent, Resolve, Absolve, Solve
--SOMA--	SOmah	Body	Somatic cell, Somatic, Chromosomes
SOMATO--	soMAT-toe	Body	Somatoblast, Somatogenic, Somatopleure
SOMNUS	SAHMnus	Sleep	Somnambulist, Somnifacient, Somniferous
SONA--	SŌN-nah	Sound	Sonata, Sonant, Consonant, Sonorous
--SOPH--	SAHf or SŌff	Wise	Philosophy, Sophisticated, Sophomore
SOPOR--	SAHPor	Lethargic Sleep	Soporific, Soporiferous, Sopor, Soporous
--SORB	Sorb	to Suck In, Swallow	Absorb, Resorb, Adsorb, Absorption
SORC--, SORT	SORss, Sort	Draw Lot, Fate	Sorcery, Sorcerous, Sortilege
SOUF, SUFF--	SOOFf, SUFf	Under-blow, Fluff	Soufflé, Souffleed,
SPADI--	SPADih	Break, Brown Color	Spadix, Spadicious, Spadicose
SPASM	SPAZm	Draw (disem-bowel), to Rend	Spastic, Spasmodic, Spasticity
SPATHE	SPĀth	Broadsword	Spathaceous, Spathose, Spathe
SPATI, SPAC--	sPATTY, SpĀce	Space, Empty	Spatial, Spacing, Spacer
SPATUL--	sPATyool	Broad Piece	Spatula, Spatulate, Spatular, Spatulum
SPECIA--	SPESSYah	Sort	Specimen, Specific, Species

ROOT/PREFIX/SUFFIX	PRONUNCIATION	MEANING	EXAMPLES OF USAGE
SPECT--	SPECKt	Behold	Spectator, Spectacle, Spectacular
--SPECT	SPECKt	to Look	Perspective, Inspect, Prospector
SPECTR	SPECKTr	Vision, Ghost	Specter, Spectral, Spectrum
SPECTRO--	SPECKTro	Spectrum	Spectra, Spectro-heliograph, Spectrometer
SPECUL--	SPECKyool	Spy, Look, See	Speculate, Speculum, Specular
SPEL--	SPELL	Cave	Spelian, Speleology, Spelunker
--SPEND--	spEND	to Weigh Out	Expense, Suspend, Spence, Spender
--SPERM--	SPURm	Seed, Sow	Spermatozoa, Spermatophore, Carposperm
SPHAC--	sFASS (as in "pass")	Gangrene	Sphacelate, Sphacelism, Sphacelous
SPHEN--	sFEEN	a Wedge	Sphenic, Sphenodon, Sphenoid
SPHERE	SFEAR	Sphere, Ball	Spherical, Spheroidal, Hemisphere
SPHRAG--	ssFRAJ	Signet, Seal	Sphragide, Sphragistics, Sphragis
SPHYGM--	ssFIGm	Pulse	Sphygmomanometer, Sphygmoid, Sphygmus
SPIC--	sPICK	Spike, a Point	Spicate, Spiculate, Spicule, Spiculum
SPINA	sPEENah	Spine, Thorn	Spinal, Spinach, Spinode, Spine
--SPIRE	SPYer	to Breath	Inspire, Respiration, Aspirator
SPIRE	SPYr or sPEER	a Coil	Spiral, Spireme, Spiriferous, Spirillum
SPIRIT	SPEARit	Breath	Spiritual, Spirits, Spirituous

ROOT/PREFIX/SUFFIX	PRONUNCIATION	MEANING	EXAMPLES OF USAGE
SPLANCH--	sPLAN-k	Viscera	Splanchnopsis, Splanchnopleura, Splanchnic
SPLEN--	Splen	Spleen, Rancor	Splenish, Splenic, Splenectopia
SPLEND--	SPLENd	Brilliant, Shining	Splendid, Splendor, Resplendent
SPOLI--	SPŌLih	Despoil	Spoliation, Spoliate, Spolium
SPONDYL--	SPAHNdl	Vertebrae	Spondylalgia, Spondylitic, Spondylopathy
SPONGE	Spunj	Sponge	Spongiform, Spongiole, Spongelet, Spunk
SPONS--	SPAHNs	Reliable, Promise	Spouse, Sponsor, Respond, Responsible
SPONT--	SPAHnt	Free Will	Spontaneous, Spontaneity, Spontaneously
--SPOR--	Spōar	Seed, Sowing	Sporocyte, Sporocyst, Macrospore
SPORADI--	spōRADDih	Scatter	Sporadic, Sporadosiderite
SPORANG--	spōRANJ	Seed-vessel	Sporangium, Sporangiophore, Sporangiospore
SPUM--	Spyoom or Spoom	Foam	Spumescent, Spumid, Spume, Spumoni
SQUAD	Skwahd	Four	Squadron, Quad, Squad, Squadrilla, Square
SQUAL--	Skwahl	Base, Foul	Squalid, Squalor, Squalidity, Squalidly
SQUAMA	SKWAHma or SKWAYma	Scale, Thin Plate	Squamal, Squamatic, Squamula, Squamous
STAB--	Stab or STAYb	Stable, Standing	Stabilize, Stable, Establish
STADI--	STAIDih	Stand	Stade, Stadium, Stadia

ROOT/PREFIX/SUFFIX	PRONUNCIATION	MEANING	EXAMPLES OF USAGE
STAGN--	STAGn	Quiet Pool	Stagnate, Stagnation, Stagnant, Staunch
STALAGM--	stahLAGm	Drop	Stalagma, Stalagmite, Stalactiform
--STALI--	STAHLee	Constriction	Peristalsis, Stethostalic, Stalling
STAMEN--	STAYmen	Thread	Stamenate, Stamina, Stamen, Staminodium
STANNI--	STANih	Tin	Stannary, Stannic, Stannite, Stannum
--STANT, STANZ--	Stant, Stands	Abode	Stanchion, Substantial, Substantiate, Stanza
STAPES	STAYpeez	Stirrup	Stapedial, Stapedectomy, Stapediovestibular
STAPHYLO--	STAFFil-oh	Bunch of Grapes	Staphylococcus, Staphyloma, Staphyloraphy
--STASI--	STAYsih or STASSih	Standing, Still	Hemostasis, Cryostasis, Isostasy, Statue
--STAT--	Statt	Stanch, Stop	Hemostat, Thermostat, Static, Status
STATO--	STATo	Standing	Statoblast, Statoscope, Stator, Statute
STAURO--	STARoh	Cross	Staurolite, Staurolitic, Staurophyll
STEAR--	STEER	Fat, Oil	Stearic, Stearate, Steatogenous, Steatoma
STEGO--	STEGgo	Roof, Cover	Stegomyia, Stegosaurus, Stegobium
STELLA--	STELL-lah	Star	Constellation, Stellar, Stellate
STENO--	STEN-no	Narrow	Stenographer, Stenochoric, Stenophylous
STERCO--	STURK-oh	Feces	Stercolith, Stercoroma, Stercobilin

ROOT/PREFIX/SUFFIX	PRONUNCIATION	MEANING	EXAMPLES OF USAGE
STEREO--	STEHRee-oh	Solid, and with Three Dimensions	Stereophonic, Stereognosis, Stereoisomer
STERNUM	STURNum	Breast	Sternalgia, Sternodymia, Sternoid
STETHO--	STETH-oh	Breast	Stethoscope, Stethometer, Stethoscopy
STHENO	ssTHEN-oh (almost like "THIN-oh")	Strength	Asthenia, Asthenolith, Sthenia, Sthenic
STIBI, STIMMI--	STIBih	Antimony	Stibine, Stibium, Stibnite
STICH	Stick	Row, a Line	Stichic, Stichometry, Stichomythia
--STIGMA--	STIGmah	Stain, Spot	Stigmatize, Stigmatic, Astigmatism
--STILL--	Still	to Drip, Drop	Instill, Distillate, Stillicidium
STIMUL--	STIMyool	Goad, to Prick	Stimulate, Stimuli, Stimulation
STIPE	STIP (sometimes Staip)	a Branch, Stem	Stipel, Stipellate, Stipes, Stipitate
STIPEND	STAIPend	Tax	Stipendium, Stipendiary, Stipend
STIPULA	STIPyoo-lah	Agree, Bargain	Stipulate, Stipulation, Stipulated
STOICH--	STOYk	Step, Go	Stoichiology, Stoichiometry, Stoichiological
STOMA	STOW̄-mah	Mouth	Stomatic, Stomatopod, Stomach, Stomata
STRABIS--	STRABBiss	to Squint	Strabismus, Strabotomy, Strabismometer
STRAIT	Straight	Bind Tightly	Strain, Straits, Straitjacket
STRANG--	STRAYing	Halter, Twisted	Strangle, Strangulated, Strangury
STRATEG--	STRATtej	Act of a General	Strategy, Strategic, Strategem

ROOT/PREFIX/SUFFIX	PRONUNCIATION	MEANING	EXAMPLES OF USAGE
STRATI--	STRATtee	Layered	Stratus, Stratiform, Stratocumulus
STRENU--	STRENyoo	Active	Strenuous, Strenuously, Strenuousness
STREPTO--	STREP-toe	Twisted	Streptococcus, Streptomycin, Streptomycetes
STRIA--	sTRYah or STREEah	Furrow	Striae, Striated, Striation, Strick
STRICT	sTRICKED	Bind	Stricture, Restriction, Constriction
STRID--	sTRID	Grating Sound	Strident, Stridor, Stridulum
STRIG--	Strig	Scraper	Strigilation, Strigilose, Strigose
STROB--	stRŌBE	Twisting, Turn	Strobic, Strobile, Stroboscope
STROMA--	stRŌMA	Bed, Bed Cover	Stromatic, Stromatolite, Stromatin
STRONGYL--	STRAHNJisle	Round	Strongylosis, Strongylidae, Strongyloid
--STROPH--	STRŌff	Turn	Apostrophe, Strophiole, Strophulus
--STRU	STROO	Build	Construe, Construct, Restructure, Obstruct
--STRUCTU--	STRUKToo	Build	Construction, Destruction, Destroy, Structure
STRUTHIO--	STROOthee-oh	Sparrow, Ostrich	Struthiones, Struthionidae, Struthiform
STRYCHN--	STRICKn	a Kind of Plant: Nightshade, Belladonna	Strychnine, Strychnia, Strychninism
STUDE, STUDI--	STOODeh, STOODih	Be Diligent, Zeal	Student, Studio, Studied, Studious
STULT--	STUHlt	Foolish	Stultify, Stultification, Stolid

ROOT/PREFIX/SUFFIX	PRONUNCIATION	MEANING	EXAMPLES OF USAGE
STUPE--	STEWp or Stoop	Make Astonished	Stupefaction, Stupid, Stupendous, Stupefy
STYL--	Stile	Stake, Column	Stylus, Styloid, Styolite
STYPTIC	STIPtick	Contract	Stypsis, Styptic, Styptic pencil
STYRE--	sTIRE	Fragrant Resin	Styracaceous, Styrene, Styrole
SUAS, SUAV--	SWAYss, SWAHv	Sweet	Persuade, Suave, Suasion, Dissuade
SUB--	Sub	Under, Lower	Substitute, Subtract, Subordinate, Substance
SUBLIM--	subBLIME	High, Lofty	Sublimate, Sublime, Sublimity
SUCCIN--	SUCKsin	Amber	Succinate, Succinylcholine, Succinic
SUCCINCT	suckSEENKT (sometimes sooSEENkt)	Below-gird	Succinctly, Succinctorium, Succinctness
SUCCU--	SUCKyoo	Juice	Succulent, Succulous, Succulence
SUCCUB--	SUCK-cube	Strumpet, Sink Down	Succumb, Succubus, Succumbent, Succubae
SUCCUS--	suhKUSS	Shaking, Shook	Succusion, Successive, Succussation
SUDA, SUDO--	SYOODah, SYOOD-oh	Sweat, Perspire	Sudarium, Sudary, Sudatory, Sudatorium
SUFFER	SUFFr	Beneath-bear	Suffering, Sufferance, Sufferable
SUFFIC--	sufFĪCE	Afford	Sufficient, Sufficiency, Suffice
SUFFRAG--	SUFFrej	Assisting	Suffragan, Suffrage, Suffragist
SUFFRUT--	SUFFfruit	Bottom of Bush	Suffrutex, Suffrutescent, Suffruticose
SUGGEST	sugJEST	Under-bring	Suggestion, Suggestibility, Suggestive

ROOT/PREFIX/SUFFIX	PRONUNCIATION	MEANING	EXAMPLES OF USAGE
SULC--	Sulk	Plow, Groove	Sulcate, Sulcus, Sulcation
SULF--	Sulf	Combination with Sulfur	Sulfate, Sulfite, Sulfohydrate
SUMMA--	SUMmah	Concise, Instant	Summary, Summation, Sum, Summarize
SUMMON	SUMmun	Slightly Warn	Summons, Summoner, Summoned, Summon
SUMPT--	SUMPt	Pack-saddle	Sumpter, Sumptuous, Sumptuary
SUPER--	SOOPer	Over, Superior	Superfluous, Superficial, Supercilious
SUPIN--	SOO-pine	Put on the Back	Supine, Supinate, Supination
SUPPLI--	SUPplih	Beg	Suppliant, Supplicate, Supplication
SUPPOS--	SUPP-oss	Under-place	Supposition, Suppository, Suppose
SUPRA--	SOOPrah	Above	Supra-esophageal, Supraliminal, Suprarenal
--SURG--	SURj	Rise	Source, Surge, Insurgent, Insurrection
SURGER--	SURJer	Hand Work	Surgeon, Surgery, Surgical, Surgically
SUSPEN--	suSPEN	Under-hang	Suspend, Suspension, Suspense, Suspender
SUSURR--	suSURR or shooSURR	Whisper, Rustling	Susurrant, Susurrus, Susurrate, Susurration
SUTUR--	SOOTcher	Sew	Souter, Sutural, Sutures, Sutura
SYCO--	SIGH-kō	Fig, Lump	Syconium, Sycophant, Sycamore, Sycosis
SYLLAB--	SILLahb	Together-take	Syllabus, Syllable, Syllabic
SYLPH	SILLf	Spirit, Beetle	Sylphid, Sylphide, Silphidae, Sylphidine

ROOT/PREFIX/SUFFIX	PRONUNCIATION	MEANING	EXAMPLES OF USAGE
SYLVA--	SILLVah	Country, Woodland	Sylvan, Sylvatic, Silvaculture
SYM--	Sim	Same, Similar	Symmetry, Sympathy, Symbol
SYMBIO--	SIMMbe-oh or simBUY-oh	Dwell Together	Symbiosis, Symbiotic, Symbiotically
SYMPHON--	SIMMphone	Together-sound	Symphony, Symphonetic, Symphonious
SYMPTOM	SIMPTum	Together-fall	Symptomatic, Symptomatically, Symptomatology
SYN--	Sin	With	Synthetic, Synapse, Synchrony, Synagogue
SYNCLIN--	SIN-kline	Together-bend	Synclitic, Synclinical, Synclinorium
SYNCOP--	SIN-cope	Together-cut	Syncope, Syncopal, Syncopation, Syncoptik
SYNCRET--	SINkreet	Unite Against	Syncretism, Syncretize, Syncrisis
SYNDE--	SINDeh	Bind Together	Syndactyl, Syndesmosis, Syndetic, Syndesmoma
SYNERGE--	SINnurj	Work Together	Synergist, Synergetic, Synergida, Synergy
SYNGENE--	SINjen-ih	Together-origin	Syngamy, Syngenesis, Syngenese, Syngenesious
SYNOD	SIN-ud	Together-way	Synodic, Synodal, Synodically
SYNONYM	SIN-o-nimm	Together-name	Synonymous, Synonymatic, Synonymic
SYNOPSIS	sinNAHPsis	Together-view	Synoptic, Synoptical chart, Synopsis
SYNOVI--	sinNŌVih	Together-egg	Synovia, Synovitis, Synpelmous, Sympetalous
SYNTHES--	SINthuss	Together-place	Synthesis, Synthetic, Synthesize

ROOT/PREFIX/SUFFIX	PRONUNCIATION	MEANING	EXAMPLES OF USAGE
SYNTON--	SIN-tone	Harmony	Syntonic, Syntonize, Syntony, Syntonical
SYRIN--	SURrin	Tube, Pipe	Syringe, Syrinx, Syringa, Syringomyelia
SYRPH--	SIRf	Gnat	Syrphidae, Syrphian, Syrphid, Syrphus
SYSTEM	SISTum	Together-stand	Systematic, Systemic, Systematist
SYZYGY	SIZih-jee	Yoked Together	Syzogetic, Syzygy, Syzagetically
TABAN--	TABban or taBAN	Horsefly	Tabanus, Tabanidae, Tabanid, Tabanomorpha
TABE--	TAYbe	Wasting Away	Tabefaction, Tabes, Tabescent, Tabetiform
TABUL, TABL--	TAByool, TAYBl	Board	Table, Tabulate, Tableau, Tabloid
TACH--	Tack	Speed, Swift	Tachometer, Tachycardia, Tachypnia
TACIT	TASSit	Be Silent	Taciturn, Tacitly, Tacitness
TACT--	Tackt	Touch	Tactile, Tactometer, Tactual, Tact
TAEN, TENIA--	TAEEn, TEENYah	Ribbon, Stretch	Taeniacide, Taenidium, Taenifuge
TALI, TALO--	TALih, Tallow	Ankle	Taliped, Talipes, Talocalcaneal, Talus
TANG--	TANj	Touch, Feel	Tangible, Tangent, Tango, Intangible
TAPE--	TAYp, TAPeh	Carpet	Tapestry, Tapeworm, Tapetum, Tapets
TARDY, TARDI--	TARdee, TARdih	Slow	Tardigrade, Tardo, Tardy, Tardive, Tarry
--TARS--	TARss	Ankle, Foot	Tarsum, Tarsae, Metatarsal, Tarsier
TAURO--	TŌRoh or TAWRoh	Bull	Taurine, Taurus, Taurocholic, Tauriform

ROOT/PREFIX/SUFFIX	PRONUNCIATION	MEANING	EXAMPLES OF USAGE
TAUTO--	TAUGHT-oh (More correctly: tahOOT-oh)	Repeat, Same	Tautology, Tautonym, Tautochrome
TAXI--	TACKSih or TACKsee	Arrangement	Taxidermy, Taxonomy, Taxiarch, Taxonomist
--TAXI	TAXih	Movement (toward or away from)	Phototaxis, Geotaxis, Taxiway
TECHN--	TEKneh	Art, Skill	Technician, Technological, Technique
TECTON--	TEXton	Carpenter	Tectonics, Tectological, Tectonosphere
TEGMEN--	TEGmen	Cover	Tegula, Integument, Tegulum, Tegular
--TEINO	TEEN-oh or TEN-no	Stretch	Protasis, Protend, Protensive
TELE--	TELLeh	Far, Long Distance	Telephone, Telegram, Telemetry, Telegony
TELEO--	TELLEE-oh or TEELEE-oh	End, Complete	Teleophase, Teleology, Teleost
TELLU--	TELLyoo or TELLoo	Earth	Tellurian, Telluric, Telluride, Tellus
TEMP--	TEMp	Time	Temper, Temperament, Temperate, Tempest
--TEMPT, TENTO--	TEMPt, TENt-oh	Touch, Try	Taunt, Attempt, Contempt, Tempting
TEN--	Ten	to Hold Fast	Tenaculum, Tenacious, Tenant, Tenet
TENDO--	TENdough	Stretch, Sinew	Tendon, Tendril, Tendinous, Tent
TENEBR--	TENeb-rah or tehNEEBrah	Gloomy, Darkness	Tenebrionidae, Tenebrific, Tenebrous
TENS--	TENss	Tension	Tense, Tensile, Tensiometer, Intense
TENU--	TENyoo	Thin	Tenuous, Tenuity, Tenuis
TEPE, TEPI--	TEPeh, TEPih	Warm	Tepify, Tepefaction, Tepid, Tephrite

ROOT/PREFIX/SUFFIX	PRONUNCIATION	MEANING	EXAMPLES OF USAGE
TER--	Tur	Thrice	Tercentennial, Tercet, Tertial, Tertiary
TERATO--	TERaht-oh	Monster	Teratism, Teratogen, Teratoid, Teratoma
TERG--	TURg or TEHRg	the Back	Tergum, Tergiversate, Tergiversation
TERM--	TURm	Boundary	Terminal, Terminate, Terminus
TERN	TURn	by Threes	Ternary, Ternion, Ternate, Ternately
TERRA--	TERRah	Earth	Terrarium, Terrestrial, Terrain
TERRI--	TERRih	Frighten, Terrify	Terrible, Terrify, Terror, Terrorist
TERTI--	TERsh or TEHRtih	Third, Thrice	Tertian, Tertiary, Terza, Tertial
TESSELL, TESSA--	TESSel, TESSah	Four-squared, Checkerboard	Tessellate, Tessera, Tessellated
TEST--	Test	Pot, Head, Shell	Test tube, Test, Testa, Testudinal, Testy
TETAN--	TETTan	Rigid	Tetanize, Tetanus, Tetanization, Tetany
TETRA--	TETtrah	Four, Fourfold	Tetrahedron, Tetrad, Tetramerous
TEXT--	Text	to Weave	Textile, Texture, Textbook, Textual
THALAM--	THALm	Chamber	Thalamus, Thalamencephalon, Thalamic
--THALAS--	THALL-us	Sea	Thalassemia, Thalasography, Thalassic
THALLO--	THALL-oh	Spring Flowers	Thallophyte, Thalloid, Thallus
--THANA--	THANNah	Death	Euthanasia, Thanatopsis, Thanatos

ROOT/PREFIX/SUFFIX	PRONUNCIATION	MEANING	EXAMPLES OF USAGE
THAUMATO--	THAWmaht-oh	Wonderful Thing	Thaumatrope, Thaumaturge, Thaumatology
THEA--	THEEah	God	Thearchy, Theanthropis, Theanthropic
--THEC--	THEEK	Bag, Sack	Oötheca, Spermatheca, Theca
THELE--	THEE-lee	Nipple, Teat	Thelitis, Thelorrhagia, Thelegenic
THEO--	THEE-oh	God	Theology, Theocentric, Theocracy, Atheist
THEOR--	THEEor	Look At	Theories, Theorem, Theoretical
THERAP--	THEHRahp	Serve	Therapy, Therapeutic, Therapist
THERIA	THEERee-ah	Wild Beast	Theriaca, Theriogenology, Theriomorphic
THERM--	Thurm	Hot, Heat	Thermometer, Thermal, Thermolysis
THES, THET--	THEEss, THEHT	Place	Synthetic, Synthesis, Thesis, Thetic
THIGM--	THIGm	Touch	Thigmotropism, Thigmotaxis, Thigmotropic
THIO--	THIGH-oh	Brimstone	Thioacetate, Thiocyanate, Thionic
THORAC--	THORAss	the Chest	Thorax, Thoracic, Thoracentesis
THROMBO--	THRAHMB-oh or THRŌMB-oh	Clot, Thicken	Thrombin, Thrombocyst, Thromboid
THYMO--	THIGHmō	Mind, Spirit	Thymocyte, Thymolysis, Thymus
--THYRO--	THIGHro	Oblong Shield	Thyroid, Hypothyroidism, Thyroiditis
TINCT--	TEENkt or TINKt	Stain	Tincture, Tinct, Tinctorial, Tingible
TINEA	TINnee-ah	Worm, Moth	Tineal, Tineidae, Tineoid

ROOT/PREFIX/SUFFIX	PRONUNCIATION	MEANING	EXAMPLES OF USAGE
TINNI--	tinNIGH	Ringing	Tinnitus, Tingle, Tintinnabulation
TITAN	TYEtan	Giant	Titanic, Titanium, Titanosilicate
TOCO--, TOKO--	TOE-kō	Bride, Obstetrics	Tocology, Tocometer, Tocher
TOLERAT--	TAHLer-ut	Bearable	Tolerant, Tolerate, Toleration
--TOM	TOEm	Cut	Atom, Diatom, Epitome, Microtome
--TOMY	TOE-me or TUM-me	Cutting	Anatomy, Tracheotomy, Hysterectomy
--TONO--	TOE-no or TAHN-oh	Stretch, Tension	Isotonic, Hypertonic, Tonic, Tonus
TONSO--	TAHNso	Barber, Shearing	Tonsorial, Tonsure, Tonsured
TOPIC--	TAHPik or TOEpik	of a Place	Topical, Topically, Topic
TOPO--	TOE-po	Surface	Topographical, Topography, Topectomy
TORPID	TORPid	Be Numb, Stiff	Sturdy, Torpedo, Torpid, Torpor
TORQU--	TORk	Collar, Twist	Torquate, Torque, Torsion, Torsade, Torte
TORRE--	TORreh	Parch, Roast	Torrent, Torrential, Torrefy, Torrid
TORSI--	TORsih	Twist	Torsion, Torsive, Tortuous, Torsiometer
TORU--	TORoo	Elevation, Swelling	Torus, Torulose, Tore, Torulus
TOTUS	TOTE-us	All	Total, Totality, Totipalmate, Totipotency
TOXI--	TAHKsih	Poison	Toxiglossa, Toxicology, Toxin, Toxic
TRABE--	TRAHb	Beam, Bar	Trabecula, Trabeculate, Trabecular

ROOT/PREFIX/SUFFIX	PRONUNCIATION	MEANING	EXAMPLES OF USAGE
TRACE--	TRAYss	Go, Draw, Follow	Tracer, Tractor, Traction, Trace
TRACH--	TRAYk	Windpipe	Trachea, Tracheal, Tracheitis, Tracheostomy
TRACHO, TRACHY	TRAHK-oh, TRAHKee	Rough	Trachoma, Trachytic, Trachyte
TRADUCE	trahDOOSS or trahDYOOSS	to Lead Over	Traducer, Traduction, Traducianism
TRAG--	tRAG	a Goat	Tragacanth, Tragedy, Tragus
TRANQUIL	TRANkwil	Quiet	Tranquilizer, Tranquility, Tranquilize
TRANS--	TRANss	Across	Transfer, Translocation, Transpose
TRANSFER	TRANss-fur	Across-bear	Transferred, Transference, Transferable
TRAPEZ--	trahPEEZ	Four-footed Bench	Trapeze, Trapezium, Trapezoid
TRAUMA	TRAWmah or TRAH-oo-ma	Wound	Traumatic, Traumatize, Traumatropism
TREA, TRAIT	TREEah, TRAYt	Give Over	Treason, Traitor, Traitorous, Treasonous
TREMEN--	TREHMen or TREEMen	Tremble	Tremendous, Tremulous, Tremens, Trepidation
TRENCH, TRUNC--	TRENch, TRUNk	Cut	Trenchant, Trencher, Entrenched
TRI--	Try	Three	Triceratops, Triangle, Tricuspid
TRIBU--	TRIBbyoo	Press, Tribe	Tribune, Tribulation, Tribunal
TRICH--	Trick or TRYk	Hair	Trichinid, Trichinosis, Trichite
TRICHOT--	TRY-kaht	Three Cutting	Trichotomy, Trichotomic, Trichotomous

ROOT/PREFIX/SUFFIX	PRONUNCIATION	MEANING	EXAMPLES OF USAGE
TRICHROM--	TRY-krōm	Three Colors	Trichroic, Trichromatic, Trichromic
TRICLIN--	TRY-klin	Leaning Three Ways	Triclinic, Triclinose, Triclinium
TRIFOLI--	TRYfoal-ee	Three Leaves	Trifoliate, Trifolium, Trifoliolate
TRIFORM--	TRYform	Three Forms, Three Shapes	Triform, Trifurcate, Trifurcation
TRIGLYPH	TRYgliff	Three Carvings	Triglyphic, Triglyphal, Triglyph
TRIGON--	TRYgahn	Triangled	Trigonal, Trigonometry, Trigonometric
TRILOBE	TRYlōb	Three Lobes	Trilobite, Trilobate, Trilocular
TRINI--	TRINih	Three	Trinity, Trinitrate, Trinitrotoluene
TRIPOD	TRYpahd	Three-footed	Tripodal, Tripody, Trivet, Tripos
TRIRAD--	TRYrad	Three-cornered	Triradial, Triradiate, Triquetrous
TRIT--	TRYt or tRICH	Rub, Worn, Frayed	Trite, Trituration, Triturate, Triturator
TRIVIAL	TRIVee-al	at Cross-roads, Common	Trivium, Trivial, Triviality
TROCHA--	TRŌ-kah	Runner, to Run	Trochanter, Trochelminth, Trochilus
TROCHL--	TRŌ-kl or TRAHK-l	Pulley, Wheel	Trochlea, Trochoid, Trochophore
TROGLOD--	TRAHGlud or TRŌGlud	Hermet, Enter a Hole	Troglodyte, Troglodytic, Troglotrema
TROPAEO--	trōPAYEE-oh	Trophy, Brightly-colored	Tropaeolin, Tropaeolum, Tropaeum, Tropaeon
--TROPE--	tROPE	A "Turn of Speech"	Trope, Tropology, Tropical

ROOT/PREFIX/SUFFIX	PRONUNCIATION	MEANING	EXAMPLES OF USAGE
TROPHO--	TRŌfo	Nourish, Feed	Trophic, Trophoblast, Atrophy
TROPO--	TRŌ-po	Defeat (of enemy)	Trophy, Tropaion, Tropaeolum
--TROPOS--	TRŌPpos	Turning toward	Tropophyte, Geotropic, Phototropism
TRUB--	tRUB	Crowd, Disturbance	Trouble, Troublesome, Troublous
TRUCULENT	TRUCKyoo-lunt	Fierce	Truculence, Truculently
--TRUD	TROOd	Thrust	Extrude, Protrude, Intrude, Protrusive
TRUNC--	Trunk	Maimed	Truncate, Truncheon, Truncated
TRYPAN--	TRIPpan	Borer	Trepan, Trypanosome, Trypanosomiasis
TUBER	TYOOBer	Swelling	Protuberant, Truffle, Tuberculosis
TUBU--	TOObyoo or TYOOByool	a Tube	Tubercle, Tuberous, Tubal, Tuba
TUIT--	TOOit or TYOOit	Defend	Tutor, Tuition, Tutelage, Tutelary
TUMU--	TOOMur or TYOOMor	to Swell	Tumor, Tumescent, Tumifacient, Tumid
TUNIC	TYOONik	a Loose Blouse	Tunicate, Tunicle, Tunica, Tunic
--TURB--	Turb	a Mob, Crowd	Troop, Disturb, Turbulent, Turbid
TURBIN--	TUR-bine (also TURbin)	Whirlwind, Spinning, Toplike, Cone	Turbine, Turbinate, Turbination, Turbinal
TURGID	TURJid	Swell	Turgescent, Turgent, Turgidity, Turgor
TURRE, TURRI--	TURreh, TURrih	Tower	Turret, Turriculate, Turrical, Turrilite

ROOT/PREFIX/SUFFIX	PRONUNCIATION	MEANING	EXAMPLES OF USAGE
TYMPAN--	TIMpan	Drum	Tympanic membrane, Tympany, Tympanum
--TYP--	Tip or Type	Style, Print	Atypical, Type, Typography, Prototype
TYPH--	TIFF or TYEff	Blind, Closed	Typhlosis, Typhoid, Typhlitis
TYRAN--	TYEran or TEERan	Master	Tyrant, Tyranny, *Tyrannosaurus rex*
TYROS--	TYE-ross	Cheese	Tyramine, Tyrogenous, Tyroid, Tyrosine
UBIQUE--	yooBIK or youBEEK	Everywhere	Ubiquity, Ubiquitous, Ubiquitist
--UDUS, UDOM--	YOODus, YOODum	Moist	Udometer, Udometric, Humid, Humidity
ULC--	ULss	Ulcer	Ulcerate, Ulceration, Ulcerous
--ULE	YOOl	Small, Diminutive	Granule, Tubule, Vestibule, Module
ULNA	ULNah	the Elbow	Ulnar, Ulnad, Ulnaris, Ulnoradial
ULTE, ULTI--	ULTeh, ULTih	Beyond	Ultimate, Ulterior, Ultima, Ultimatum
ULTRA--	ULTrah	Extreme	Ultramarine, Ultramontane, Ultraviolet
ULUL--	OOLool	Howl	Ululant, Ululation, Ululate
--UMBRA--	OOMBrah or UMBrah	Shade	Umbrella, Umbrage, Penumbra
UNCI--	UNsih	Hook-shaped	Uncus, Uncinate, Uncial, Unsiform
UNCTIO--	UNKtee-oh or UNKshee-oh	to Annoint, Oil	Unction, Uncture, Unctuous
UNDA, UNDU	UNDah, UNDyoo	a Wave	Undulate, Inundate, Undulation, Undeé
UNDECIN--	OONdeh-sin	Eleven	Undecennial, Undecagon, Undecylenic acid

ROOT/PREFIX/SUFFIX	PRONUNCIATION	MEANING	EXAMPLES OF USAGE
UNGU--	UNGoo or UNGwah	Nail, Hoof	Ungulate, Ungula, Unguis, Ungual
UNI--	YOONih or OONih	One, Single	Uniaxial, Unicorn, Unicycle, Union
UNIP--	YOOnip	One-produce, One-foot	Uniparous, Unipetalus, Uniped, Unipolar
URANI--	yooRAINee	Heavenly, Celestial	Urania, Uranian, Uranium, Uranus
URB--	Urb	City	Urban, Suburbs, Urbane, Urbis
URE--	YOOr	Urine	Enuresis, Urease, Urea, Urethra
URO--	YOOR-oh	Tail	Urochordate, Uropygial, Uropygidium
URSA--	URSah	a She-bear	Ursa Major, Ursine, Ursula, Ursiform
URTICA--	URTik-ah	Nettle, Burn	Urticaceous, Urticaria, Urticating
USTI, USTU--	UStee, UStoo	Scorch, Burn	Ustion, Ustulate, Ustulation
UTERIN--	YOOTERin	the Womb	Uterus, Uterotomy, Uteroplacental
UTI	YOOTih	Fit for Use	Utensil, Utilize, Utility
UTOPIA	yooTOEpee-ah	No Place	Utopian, Utopianism, Utopia
UTRICUL--	YOOtrick-l	Skin Bag	Utriculate, Utricular, Utriculitis
UVUL--	YOOVah (sometimes OOVah)	Grape	Uvulitis, Uvulatomy, Uvular, Uvulus
UXOR	UKSsor	Wife	Uxorious, Uxoricide, Uxorial, Et ux
UZU, USU--	Use-You, YOOsoo	Use	Usual, Usury, Usufruct, Usurp
VACAN--	VAYkan	Empty, Quit	Vacant, Vacancy, Vacate, Vacation

ROOT/PREFIX/SUFFIX	PRONUNCIATION	MEANING	EXAMPLES OF USAGE
VACC--	VAKs (sometimes VAHKs)	Cow	Vaccine, Vaccinal, Vaccination
VACILL--	Vassal	Waver, Fluctuate	Vacillate, Vacillation, Vacillatory
VACU--	VACKyoo	Hole, Cavity	Vacuole, Vacuous, Vacuity, Vacuum
--VADE, VADO	VAYD, VAHD-oh	to Go	Invade, Invasion, Evade, Invader
VAGA--	VAGGah	to Wander	Vagabond, Vagary, Vague, Vagrant
VAGIN--	VAJjin	Sheath, Cup, Inward	Vaginal, Invaginate, Vulvovaginitis
VAIN--	VAYn	Empty	Vain, Vanity, Vanish, Vanishing
--VALE--	VAYle or VALeh	Value, Worth	Equivalent, Valence, Trivalent, Invalid
VALEO--	VALLEY-oh or VAYlee-oh	to be Strong	Valiant, Valid, Invalid, Valiancy
VALL--	VALL	Wall	Vallation, Vallecula, Vallecular
VALV	VALLv	Door	Valve, Valvate, Valvulitis
--VARICUS	VARIKus	Straddle	Prevaricate, Divaricate, Prevarication
VARIO--	VARY-oh	Change, Bent	Varicella, Varicose veins, Variant
VASO--	VASS-oh	(blood) Vessel	Vasodilator, Vascular, Vasomotor
VATI--	VATih	Prophet, Prophesy	Vatic, Vatican, Vaticide, Vaticinate
--VEC--	Veck	to Carry	Convection, Convey, Invected, Vector
VEGET--	VEJet	Quicken, Animate	Vegetate, Vegetable, Vegetation
VELA--	VELlah	Fleece, Palate	Velum, Velar, Veliform, Vellus

ROOT/PREFIX/SUFFIX	PRONUNCIATION	MEANING	EXAMPLES OF USAGE
VELA--	VELlah	Sail, Covering	Veil, Velamen, Velarium, Velation
VELLICAT	VELLih-kate	to Pluck, Twitch	Vellicate, Vellication, Vellicative
VELO--	VEL-lo	Speed, Swift	Velocimeter, Velocipede, Velodrome
VENA--	VEEnah	Vein	Venation, Veins, Vena cava
VENAL	VEENal	Sale, Mercenary	Vendor, Venal, Venality, Vendue
VENERAB--	VENner-ab	Revere, Ancient	Venerate, Veneration, Venerability
VENO, VENER--	VEN-no, VENer	to Hunt	Venison, Venetorial, Venatical
VENI, VENO--	VENih, VENno	Poison	Venom, Venomous, Venomization
VENTI--	VENTih	Wind, Fan	Ventilate, Ventiduct, Ventilation
VENTR--	VENTr	Little Belly	Ventricular, Ventriloquist, Ventricose
--VENU, VENIO--	VENyoo, VENyo	Come	Adventure, Venturesome, Venue, Prevent
VENUL--	VENyool	Little Vein	Venule, Venulous, Venular, Venulose
VERA--	VERah	True	Verify, Veracity, Vera causa, Verdict
VERB--	Vurb	Word	Verbal, Verbosity, Verbatim, Verbalize
VERD	VEHRd	Green	Verdant, Verdure, Verditur, Verdigris
--VERG	VERj	to Incline	Converge, Diverge, Divergent
VERM--	VURm (also VEHRm)	Worm	Vermifuge, Vermiform appendix, Vermin
VERNA--	VURNal (also VERRnul)	Spring, Green	Vernal equinox, Vernation, Vernicose

ROOT/PREFIX/SUFFIX	PRONUNCIATION	MEANING	EXAMPLES OF USAGE
VERRUC--	verROOK	Wart	Verruca, Verrucano, Verrucose
--VERS--	Verse	Turn	Reverse, Obverse, Universe, Versatile
--VERT--	VURt	to Change, Turn	Invert, Convert, Divert, Vertex
VERTEBR--	VERTeb-r	Joint	Vertebrate, Vertebrae, Invertebrate
VESIC--	VESSik	Blister	Vesicant, Vesiculator, Vesiculating
VESP--	VESp	Wasp	Vespid, Vespulidae, Vespiary
VESPER	VESPr	Evening	Vespers, Vesperal, Vespertilian
--VEST--	Vest	Garment	Invest, Vestment, Vestry, Divest
VESTIB--	VESTib	Cavity, Entrance	Vestibule, Vestibulotomy, Vestibular
VESTIG--	VEStij	Trace, Footstep	Investigate, Vestige, Vestigial
VETER, VETUS	VETTr, VEHTus	Old, Old Soldier	Inveterate, Veteran, Inveteracy
VEX	VECKs	to Shake	Vexatious, Vexing, Vexed
VIA--	VIEah or VEEah	Way, Road	Viaduct, Viatic, Viagraph, Devious
VIBR--	VIEb-r (sometimes VIBr)	Shake, Vibrate	Vibrant, Vibraculum, Vibrato, Vibrissa
VICAR	VIKKr	Change Place	Vicarious, Vicar, Viceroy, Vicarage
VICIA	VISshe-ah or VICE-ee-ah	Vetch, Bean	Vicine, Vetch, Vetchling, Viscia
VILL--	Vill	Shaggy Hair	Villus, Villous, Villi, Villitis
VIM--	Vim	Twig, Plait	Viminal, Viminous, Vimin
VIN--	Vin	Wine, Vinegar	Vintage, Vinegerone, Vinegar, Vinic

ROOT/PREFIX/SUFFIX	PRONUNCIATION	MEANING	EXAMPLES OF USAGE
VINC	VINss	Conquer	Vanquish, Invincible, Convince
VIRG--	VURj or VURg	Slender as a Rod	Virgate, Virgulate, Virga, Virgin
VIRID--	VEERid	Greenness	Viridian, Viridescent, Viridity
VIRIL--	VEERul	Man, Courage	Virtue, Virile, Virility, Virtual
VIRU--	VEERoo or VIEruh	Slime, Ooze	Virus, Virology, Virulent, Virulence
VIS--	VISS or Vīze	Look	Visor, Visage, Vis-à-vis, Vision
VISCO--	VISk-oh	Birdlime	Viscid, Viscosity, Viscose
--VIT, VIV--	vIT, vIV	Live, Life	Vivacious, Viable, Revive, Vivid, Vitamin
VITA--	VIEtah	Life	Vitamin, Vital, Vitality, Revive
VITR--	VITTr	Glass	Vitreous, Vitrify, Vitriform, In vitro
VITU--	VITtoo or VIEtoo	Fault	Vituperate, Vitiate, Vitiation
VIV--	Viv or VIEv	Live, to Bear	Vivipara, Viviparous, Vivacious, Revive
--VOC--	VŌk	Call	Invoke, Revoke, Vocal, Vocation
VOCIF--	voSIFF	Loud Voice	Vociferate, Vociferance, Vociferant
--VOLI, VOLO--	Volley, VŌl-oh	Will	Volition, Malevolent, Volitive, Volunteer
VOLO, VOLA--	VŌlo, VŌlah	to Fly	Volant, Volatile, Volplane, Volery
VOLT--	Volt	Turn	Volti, Revolt, Voltigeur, Voluble
--VOLV	VAHLv	Roll	Convolution, Devolve, Involute, Involve

ROOT/PREFIX/SUFFIX	PRONUNCIATION	MEANING	EXAMPLES OF USAGE
--VORE--	VOReh	Eating	Voracious, Carnivore, Herbivore, Omnivore
VOT--	Vote	to Vow	Votive, Vote, Votary, Devoted
VULN--	VULn	Attack, Wound	Invulnerable, Vulnerable, Vulnerability
VULPE--	VUHLPeh	Fox	Vulpine, Vulpecular, Vulpicide
XANTH--	Zanth	Yellow	Xanthine, Xanthophane, Xanthophyll
XENO--	ZEEN-oh	Stranger, Guest	Xenophobia, Xenolith, Xenogamy
XERO--	ZEHR-oh	Dry	Xerography, Xeroderma, Xerophthalmia, Serene
XIPHI--	ZIFFEY	Shaped Like a Sword	Xiphoid, Xiphisternum, Xiphoiditis
XYLO--	ZAIL-oh	Wood	Xylem, Xylophone, Xylene, Xylem ray
XYS--	Ziss	File, Scrape	Xyst, Xyster, Xysma
ZEO--	ZEE-oh	Boil	Zeolite, Zeoscope, Zeoscopic
ZO--	Zō	Animal	Zoo, Protozoa, Zodiac, Zoöphyte
ZONE	Zone	Girdle, Belt	Zonary, Zonate, Zoned, Temperate zone
ZYGO--	zEYE-go	Yoke	Zygote, Zygospore, Heterozygous
--ZYME--	zEYEm	Leaven, Ferment	Enzyme, Zymogen, Zymometer, Zymotic